Microstructure Atlas of P91 Steel

Rajat K. Roy · Anil K. Das · Avijit Kumar Metya ·
Avijit Mondal · Ashis Kumar Panda · M. Ghosh ·
Satish Chand · Sarmishtha Palit Sagar ·
Swapan Kumar Das · Amit Chhabra ·
Swaminathan Jaganathan · Amitava Mitra

Microstructure Atlas of P91 Steel

Springer

Rajat K. Roy
CSIR-National Metallurgical Laboratory
Jharkhand, India

Avijit Kumar Metya
CSIR-National Metallurgical Laboratory
Jharkhand, India

Ashis Kumar Panda
CSIR-National Metallurgical Laboratory
Jharkhand, India

Satish Chand
NTPC Energy Technology Research Alliance
Greater Noida, Uttar Pradesh, India

Swapan Kumar Das
CSIR-National Metallurgical Laboratory
Jharkhand, India

Swaminathan Jaganathan
CSIR-National Metallurgical Laboratory
Jharkhand, India

Anil K. Das
NTPC Energy Technology Research Alliance
Greater Noida, Uttar Pradesh, India

Avijit Mondal
NTPC Energy Technology Research Alliance
Greater Noida, Uttar Pradesh, India

M. Ghosh
CSIR-National Metallurgical Laboratory
Jharkhand, India

Sarmishtha Palit Sagar
CSIR-National Metallurgical Laboratory
Jharkhand, India

Amit Chhabra
NTPC Energy Technology Research Alliance
Greater Noida, Uttar Pradesh, India

Amitava Mitra
CSIR-National Metallurgical Laboratory
Jharkhand, India

ISBN 978-981-19-9561-3 ISBN 978-981-19-9562-0 (eBook)
https://doi.org/10.1007/978-981-19-9562-0

This Springer imprint is published by the registered company Springer Nature Singapore Pte Ltd.
The registered company address is: 152 Beach Road, #21-01/04 Gateway East, Singapore 189721, Singapore

This work is prepared by CSIR-National Metallurgical Laboratory, Jamshedpur, India; in-collaboration with NTPC-NETRA, Greater Noida, India

Acknowledgements

The authors take this opportunity to express their gratitude to heads of CSIR-NML, IPU-CSIR, and NTPC-NETRA for their kind permission to publish this research work and wish to record their heartfelt thanks to Shri Ujjwal Kanti Bhattacharya, Director (Project), NTPC Ltd. and Mr. Shaswattam, Chief General Manager (CGM), NTPC-NETRA for their continuous guidance, active interest, untiring co-operation and financial support to execute the proposed investigation.

Special thanks are extended to Mr. Binod Kalindi and Mr. Ankur Chatterjee for mechanical tests, Mr. Anuj Macchua and Mr. Sadashiv Kalindi for metallographic investigation, Dr. Bhupeswar Mahato for TEM investigation, and Dr. Siuli Dutta for scanning electron microscopy during the course of investigation. Finally, the unconditional support from other staff members of CSIR-NML at various stages of investigation in the needy hours is also gratefully acknowledged. The team is thankful to NTPC Sites (Kahalgaon and Barh) for providing the samples of P91 for the experimentations.

Contents

1 Metallurgy of P91 Steel 1
References 3

2 As-Received Pipe Details 5
2.1 Dimensions: Pipe and Creep Specimen 5
2.2 Material Composition 5
2.3 Mechanical Properties 5

3 Creep Testing Under Different Conditions 9
3.1 Creep Life of P91 Steel 9
3.2 Creep Curves 10

4 Micrograph Description 13
4.1 As-Received Pipe 13
4.2 Micrographs of Creep Interrupted P91: 550 °C Temperature and 200 MPa Stress 13
4.3 Micrographs of Creep Interrupted P91: 600 °C Temperature and 140 MPa Stress 13
4.4 Micrographs of Creep Interrupted P91: 650 °C Temperature and 100 MPa Stress 13

5 Effect of Creep Exposure on Material Behaviors 95
5.1 Precipitate Size Variation 95
5.2 Hardness Variation 95
5.3 Magnetic Parameter Variation 95
5.4 Nonlinear Ultrasonic (NLU) Parameter Variation 101
5.5 Conventional Ultrasonic Parameter Variation 103

Supplements 105

About the Authors

Dr. Rajat K. Roy is currently positioned as a Senior Principal Scientist at CSIR-National Metallurgical Laboratory, Jamshedpur (India). After completing his Bachelor of Engineering and Master of Engineering in Metallurgy from Bengal Engineering College, Shibpur, Howrah (India), he obtained his Ph.D. degree in Metallurgical and Materials Engineering from Indian Institute of Technology, Kharagpur (India). His research areas are compositional design and development of nanostructured and amorphous materials through rapid solidification, metal joining (brazing), characterization of ferritic steels by electromagnetic NDE. He has authorships of 54 papers in SCI journals, 12 patents, 3 copyrights and 3 book chapters. Dr. Roy has exposures on the development of nanomaterials in Carnegie Mellon University, Pittsburgh (USA) under BOYSCAST fellowship and Korea University, Seoul (Republic of Korea) under BK21 program.

Mr. Anil K. Das is working as General Manager at NTPC Energy Technology Research Alliance (NETRA), the R&D wing of NTPC Limited (India). He is heading the Advanced Scientific Service Division there, which has labs/group related to Non-destructive Evaluation (NDE), Advanced Material Research, Creep & Fatigue, Corrosion and Environment, Analytical Chemistry, Lub & Transformer Oils, Coal & Combustion and Resin. He has done his B.Tech and M.Tech both from I.I.T. Kanpur (India). He has more than 30 years of experience in health and life assessment of power plant components. He was awarded National NDT Award in 2011 for Excellence in Application of NDE in Industries. He has filed 5 patents (2 granted) and 5 copyrights. He is life member of ISNT and member of ASNT. He is ASNT Level-3 certified in four NDT Methods (UT, MT, ET & PT) and CSWIP-Level 2 in PAUT. He is associated with more than 350 publications of reports and technical papers. His major work areas are advanced inspection and analysis tools for power plant components.

Dr. Avijit Kumar Metya has been working at CSIR-NML, Jamshedpur (India) as Principal Scientist. He obtained his B.E. (Mechanical) from Jalpaiguri Government Engineering College, West-Bengal (India); M.Tech. in Materials Engineering from BESU, Shibpur (Currently IIEST, Shibpur) Shibpur, Howrah (India) and Ph.D. from Indian Institute of Technology Madras, Chennai (India). His major research interests include Ultrasonic NDE, acoustic wave propagation through materials, mechanical behaviour of materials. He has published 16 papers in International Journals. Dr. Metya received NDT Young Scientist/Engineer Award in 2011 from Indian Society for Nondestructive Testing.

Dr. Avijit Mondal is currently working as Senior Manager at NTPC Energy Technology Research Alliance (NETRA), India. He obtained his B.E and M.Tech. from Jadavpur University, Kolkata (India) and Ph.D. from the Indian Institute of Technology, Kanpur (India). His major research area includes structure properties correlation, advanced NDE methods for condition assessment, metallurgical degradation and failure analysis of power plant components. Before joining NTPC, he was working as visiting Research Scholar at the Pennsylvania State University, USA. He was awarded Shunsaku Katoh Award for most significant contributions to scientific research in the field of Microwave and RF power at the 1st Global Congress on Microwave Energy Applications (GCEMA 2008), Japan. He was selected as one of the two participants from IIT Kanpur for SAARC-JENESYS 2009 Program by the Japanese Government. He has authored about 32 publications in international journals, conference proceedings and books. He has filed 2 patents and 4 copyrights.

Dr. Ashis Kumar Panda is working as Senior Principal Scientist in the Functional Materials (FM) group of Advanced Materials & Processes (AMP) Division of National Metallurgical laboratory, Jamshedpur (India). He obtained his Ph.D. degree in Materials Science from Indian Institute of Technology, Kharagpur (India) in the year 2002. His research interest is in the development of magnetic alloys (nanostructured, ferromagnetic shape memory, magnetocaloric) through rapid solidification, non-destructive evaluation through magnetic techniques and development of magnetic sensors. He has research experience in Slovak Academy of Science (SAS), Bratislava, Slovakia and National Institute of Materials Science (NIMS), Tsukuba, Japan under Indian National Science Academy (INSA) visiting program and CSIR-Raman Research fellowship respectively. He has about 964 publications in SCI journals and has filed 18 patents.

Dr. M. Ghosh is at present Senior Principal Scientist in Microstructural Engineering Group of Materials Engineering Division, CSIR-National Metallurgical Laboratory, Jamshedpur 831007 (India). After completing B.E. in Metallurgy from Bengal Engineering College, Shibpur, Howrah 711103 (India) and M.E. in Physical Metallurgy from BE College (Deemed University), Howrah 711103 (India), he obtained his Ph.D. in Engineering from BESU, Howrah 711103 (India). Dr. Ghosh was awarded with Indo-US Fellowship in the year of 2009 for post-doctoral research in abroad on advanced joining protocols. He has more than seventeen years of experience in research and development with a major thrust in the area of structure-property co-relation of metallic materials, transmission electron microscopy, solid state joining and failure investigation of engineering components. He has published more than 115 papers in the journals of international repute along with 15 conference proceedings.

Mr. Satish Chand is currently working as Senior Manager at NTPC Energy Technology Research Alliance (NETRA), India. He obtained his M.Sc. (Physics) from Meerut University (India). He has more than 27 years of experience in the field of health and life assessment of power plant components. His major research area includes developing procedures and methodologies for NDE inspection of power plant components, failure investigation, providing new and innovative solutions. He is ASNT Level-3 certified in 2 NDT Methods (UT & MT) and CSWIP Level 2 in PAUT. He has published more than 15 papers in in-house journal and in international conferences. He has also one patent on his name. He is associated with filing of 4 patents (2 granted) and 4 copyrights.

Dr. Sarmishtha Palit Sagar obtained her Master of Technology degree in Electronics and Communication Engineering from University of Calcutta, Kolkata, India in 1993 and her Ph.D. from Jadavpur University, Kolkata, India in 2003. She completed her post-doctoral studies at Fraunhofer Institute for Non-Destructive Testing (IzFP), Saarbreucken, Germany & Pennsylvania State University, USA. She currently holds the position of Chief Scientist & Group Leader, NDE group in CSIR-National Metallurgical Laboratory, Jamshedpur, India. She has published 60 SCI/non-SCI journal papers, 13 patents and 06 copyrights. Her current research interest includes, Non-Destructive Evaluation using Ultrasonic, Thermography and Advanced Ultrasonic, Sensor & devices for industrial applications, Signal analysis, Internet of Things (IoT) and Artificial Intelligence (AI). She is an editorial board member of the Journal of Non-destructive Evaluation. Dr. Palit Sagar received "National NDT Award in R&D" in 2005 and "National NDT award for international recognition" in 2017 and inducted as Honorary fellow of Indian Society for Non-destructive Testing (ISNT).

Dr. Swapan Kumar Das Ex-Senior Principal Scientist of CSIR-National Metallurgical Laboratory, and Professor of AcSIR was superannuated in February 2020 after 30 years of service. He obtained his M.Engg. and Ph.D. in Metallurgical & Materials Engineering from IIEST, Shibpur, Howrah 711103 (India). His areas of specialization are Physical Metallurgy, Materials Characterization and Failure Analysis. He has published more than 100 papers in reputed SCI journals.

Mr. Amit Chhabra is currently Senior Manager at the Mechanical Section, NTPC Khargone (India). Earlier, he worked in NDE, Life Management Laboratory at NTPC Energy Technology Research Alliance (NETRA, R&D wing of NTPC). He obtained his B.Tech. (Mechanical) from National Institute of Technology, Jaipur (India). His major work areas at NETRA include developing procedures and methodologies for NDE inspection of power plant components, failure investigation, providing new and innovative solutions. He is CSWIP Level 2 certified in PAUT & ToFD. He has published more than 10 papers in in-house journal and in international conferences. He has also one patent on his name.

Dr. Swaminathan Jaganathan is currently working in the position of Chief Scientist at Creep Lab., CSIR-National Metallurgical Laboratory, Jamshedpur (India). He has 28 years research experiences on creep deformation of high temperature alloys. His research interests are remaining life assessment of high temperature components and failure analysis of industrial steel components. He has published about 35 nos. research papers in various reputed journals.

Prof. Amitava Mitra is currently Visiting Professor at Department of Physics and Professor In-charge (Infrastructure), Indian Institute of Technology Jodhpur, Rajasthan (India). He did his Ph.D. from Indian Institute of Technology, Kharagpur (India) and worked about 30 years at CSIR-National Metallurgical Laboratory, Jamshedpur (India). His major areas of research interest include Magnetism & Magnetic Materials and their applications in the field of structural integrity assessment of engineering components. He published more than 150 papers in SCI journals and filed about 15 patents in India and abroad. He is the recipient of National Metallurgist Day Award: Metallurgist of the year-2006 conferred by Ministry of Steel, Government of India for his contributions in the field of Metal Science. He is also the recipient of National NDT award in the year 1997 for R&D and in the year 2000 for system development which were conferred by Indian Society for NDT. He received the Materials Research Society of India Medal in the year 2011. He is an USAID Fellow and worked at Ames Laboratory, Iowa State University, USA and also a JSPS invitation Fellow and worked at Institute of Materials Research, Tohoku University, Sendai, Japan and Iwate University, Morioka, Japan. He is a Fellow of Indian Institute of Metals, Institute of Engineers (India) and Indian Society for NDT.

Chapter 1
Metallurgy of P91 Steel

The efficiency of thermal power plants is controlled by temperature and pressure of steam. However, the operation at elevated service condition (570–600 °C, 170–230 bar) to have the benefit of higher efficiency and reduced carbon footprint requires the use of special materials to withstand the creep degradation to offer long life and ease of fabrication and maintenance. One such material known as P91 (modified 9Cr-1Mo) steel is widely used for thick section of different components in power plant due to its enhanced creep, oxidation and thermal fatigue resistance at high temperature (~600 °C). The improved creep strength results in increasing steam parameters and low CO_2 emission in atmosphere. As per ASME Boiler and Pressure Vessel Code (BPVC) Section 2A, the typical chemical composition of SA335 P91 steel is 0.08–0.12 C, 8.00–9.50 Cr, 0.85–1.05 Mo, 0.20–0.50 Si, 0.30–0.60 Mn, 0.01 S, 0.02P, 0.18–0.25V, 0.03–0.07 N, 0.40 max Ni, 0.02 max Al, 0.06–0.10 Nb, 0.01 max Ti, 0.01 max Zr, rest Fe (all in wt%).

Considering the alloy design, Chromium as a ferrite stabilizer forms carbides and provides high temperature strength along with adequate oxidation resistance. Molybdenum slows down creep deformation, reduces coarsening rate of carbide and encourages the precipitation of Laves phase. It also increases hardenability and improves thermal fatigue life. Nickel is an austenite stabilizer, which graphitizes carbide and delays the separation of carbides and increases diffusion coefficient of carbon to propel coarsening of $M_{23}C_6$. It improves toughness and strength and decreases the lower critical austenitizing temperature Ac_1. It lowers the ductile/brittle transition temperature. Vanadium is a ferrite stabilizer and carbide former and helps in retaining fine subgrains. Niobium is a carbide former, promoter of z-phase Nb(C, N) and refiner of subgrain. Titanium forms carbide and improves creep strength. Nitrogen is an austenite former, suppresses delta ferrite formation, promotes VN precipitation, increases hardenability, encourages strain aging, and lowers Ms temperature. Manganese provides weak strengthening, stabilizes austenite, delays the separation of carbides, promotes $M_{23}C_6$ coarsening, enhances M_6C precipitation, lowers Ac_1 temperature, and promotes deoxidation. Being a ferrite stabilizer, Silicon forms Laves phase, degrades mechanical properties like creep strength and toughness. However,

R. K. Roy et al., *Microstructure Atlas of P91 Steel*,
https://doi.org/10.1007/978-981-19-9562-0_1

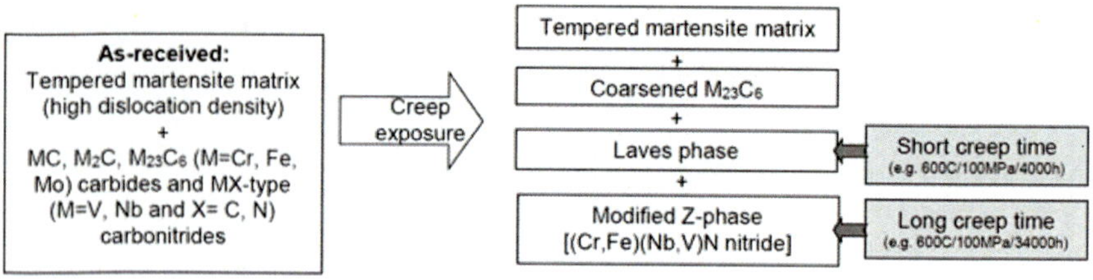

Fig. 1.1 Microstructural evolution of P91 onward creep exposure

it improves oxidation resistance. It also graphitizes carbides and steers carbon diffusivity to enhance diffusion coefficient, resulting in $M_{23}C_6$ coarsening. The addition of Aluminium graphitizes the carbide and even in small quantity, it degrades mechanical property. However, it enhances the high-temperature oxidation resistance.

The heat treatment of P91 steel corresponds with the sequence of austenitization, air cooling, and tempering treatment. After austenitization above Ac_1 temperature by dissolving most of the carbides and nitrides, the air cooling to room temperature results in fully martensitic structure with high dislocation density. Thermal treatment consequently makes the steel hard and brittle, which is relieved by tempering at temperature of 680–780 °C. During tempering, various metastable precipitates are formed, and some of them at the same time disappear, i.e., replaced by more stable precipitates. It perpetually affects high-temperature creep properties. A typical precipitation sequence in P91 steel may be as follows, where unstable primary trigonal M_7C_3 at lath boundary and dislocation forms $M_{23}C_6$ under favorable condition of temperature/stress:

$$M_3C \rightarrow M_7C_3 + M_2X \rightarrow M_{23}C_6 + MX \rightarrow M_{23}C_6 + \text{Z-phase}$$

The common precipitates in P91 steel are $M_{23}C_6$ [(Cr, Fe, Mo, W)$_{23}$(C, B)$_6$], M_2X (M=Cr, Mo and X=C, N), MX (M=V, Nb, Ta, Ti and X=C, N) carbonitrides and M_6X. Apart from them, AlN, Laves phase of type (Fe, Cr)$_2$(W, Mo) and Z-phase [Cr (V, Nb, Ta) N] are frequently observed [1–4]. Both stress and temperature play significant role in altering the microstructure, which adversely affects the mechanical properties and service life of component.

During creep deformation major changes are

- matrix grain structure/size;
- disappearance of selected primary precipitates by inter-reaction/dissolution;
- secondary precipitation of new phase;
- growth of pre-existing stable phases; and
- defect density and characteristics.

A schematic of microstructure evolution under creep deformation is illustrated in Fig. 1.1.

References

1. L. Cipolla, H.K. Danielsen, D. Venditti, P.E. Di Nunzio, J. Hald, M.A.J. Somers, Conversion of MX nitrides to Z-phase in a martensitic 12% Cr steel. Acta Mater. **58**, 669–679 (2010)
2. V. Thomas Paul, S. Saroja, M. Vijayalakshmi, Microstructural stability of modified 9Cr–1Mo steel during long term exposures at elevated temperatures. J. Nucl. Mater. **378**, 273–281 (2008)
3. M. Taneike, K. Sawada, F. Abe, Effect of carbon concentration on precipitation behavior of $M_{23}C_6$ carbides and MX carbonitrides in martensitic 9Cr steel during heat treatment. Met. Mater. Trans. A **35**, 1255–1262 (2004)
4. S. Saroja, P. Parameswaran, M. Vijayalakshmi, V.S. Raghunathan, Prediction of microstructural states in Cr-Mo steels using phase evolution diagrams. Acta Metall. Mater. **43**, 2985–3000 (1995)

Chapter 2
As-Received Pipe Details

2.1 Dimensions: Pipe and Creep Specimen

Figure 2.1.

2.2 Material Composition

Table 2.1.

2.3 Mechanical Properties

Figure 2.2 and Table 2.2.

R. K. Roy et al., *Microstructure Atlas of P91 Steel*,
https://doi.org/10.1007/978-981-19-9562-0_2

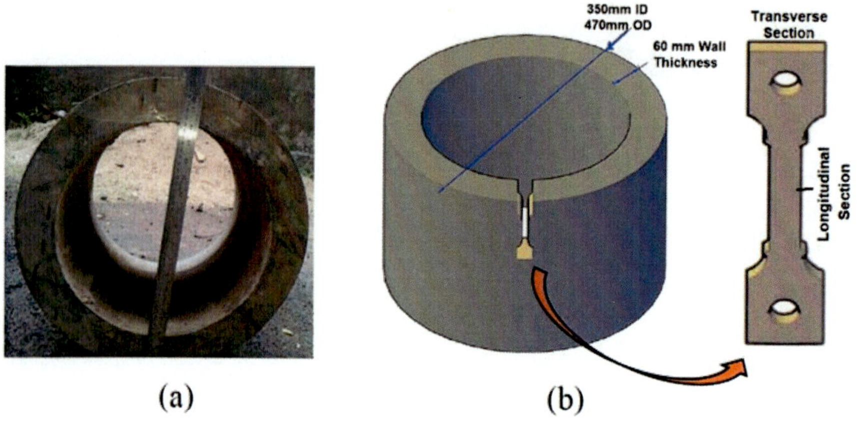

Fig. 2.1 **a** Industrially processed as-received pipe, **b** Schematic diagram of creep specimen and sectioning position from pipe for metallography and hardness measurement

Table 2.1 Chemical composition (wt%) of P91 steel

C	Si	Mn	P	S	Cr	Mo	Ni	Nb	V	Fe
0.135	0.275	0.46	0.012	0.004	8.4	0.90	0.13	0.09	0.24	rest

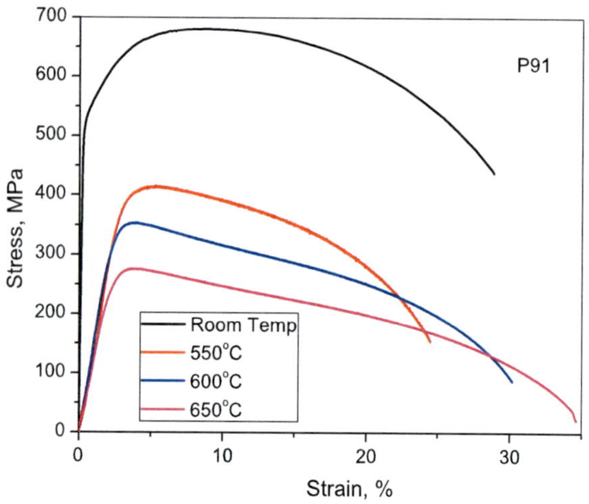

Fig. 2.2 Stress–strain curves of P91 (9Cr-1Mo) steel at room and high temperatures

Table 2.2 Mechanical properties of P91 at room temperature and high temperatures

Temp. (°C)	Hardness (HV30)	Fracture toughness, $J_{0.2}$ (kJ/m^2)	Tensile properties		
			Yield strength (MPa)	UTS (MPa)	Fracture strain (%)
Room Temp.	213.3 ± 1.2	676	525 ± 5.5	674 ± 4.4	22 ± 3.8
550	–	–	343 ± 7.8	397 ± 13.2	20 ± 7.5
600	–	–	281 ± 12.4	319 ± 16.5	25 ± 8.8
650	–	–	194 ± 4.8	243 ± 16.0	17 ± 8.7

Chapter 3
Creep Testing Under Different Conditions

3.1 Creep Life of P91 Steel

The creep life of P91 steel is determined at different temperature and stress, as mentioned in Table 3.1. It is noteworthy that the creep life varies for different samples at same temperature and stress condition. The corresponding creep curves are explained in Figs. 3.1a, 3.2a, and 3.3a, respectively.

Table 3.1 Creep life of P91 at different conditions

Creep conditions (Temp./Stress)	Sample no	Creep life (h)
550 °C/200 MPa	550 (1)	6763
	550 (2)	6815
	550 (3)	4374
600 °C/140 MPa	600 (1)	2592
	600 (2)	2598
650 °C/100 MPa	650 (1)	310
	650 (2)	614

R. K. Roy et al., *Microstructure Atlas of P91 Steel*,
https://doi.org/10.1007/978-981-19-9562-0_3

3.2 Creep Curves

Figures 3.1, 3.2 and 3.3.

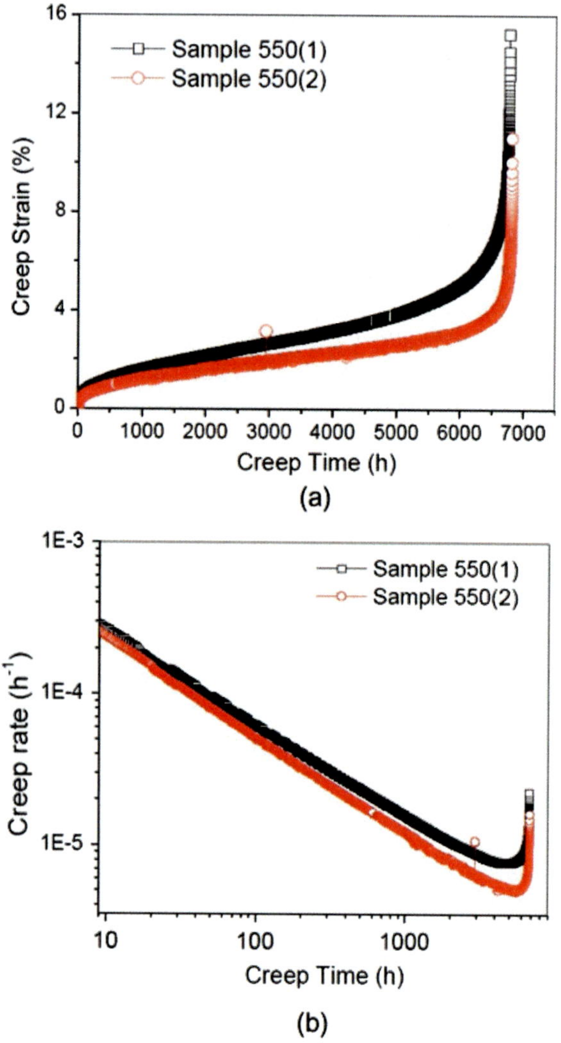

Fig. 3.1 Creep properties of P91 at 550 °C under 200 MPa stress. **a** Creep strain and **b** Creep strain rate, as a function of time

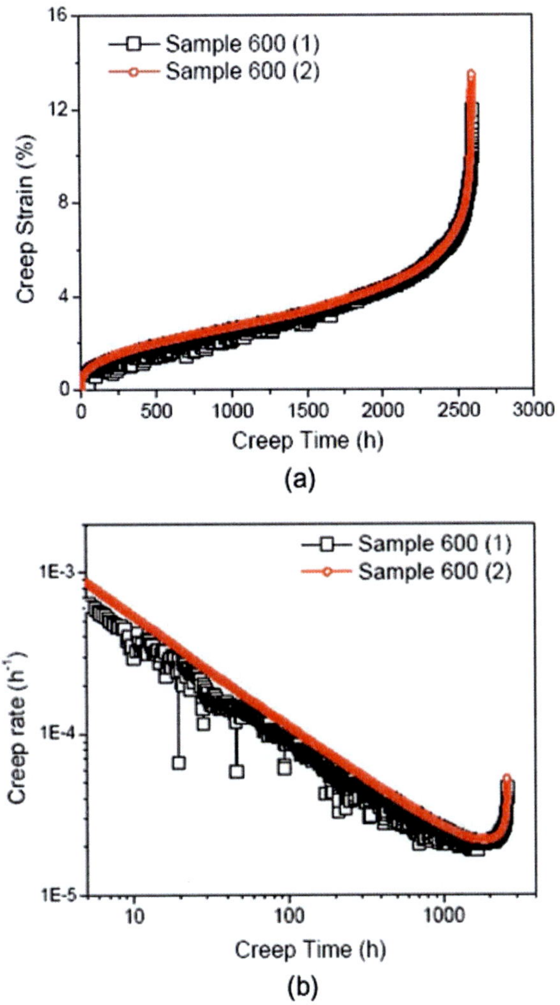

Fig. 3.2 Creep properties of P91 at 600 °C under 140 MPa stress. **a** Creep strain and **b** Creep strain rate, as a function of time

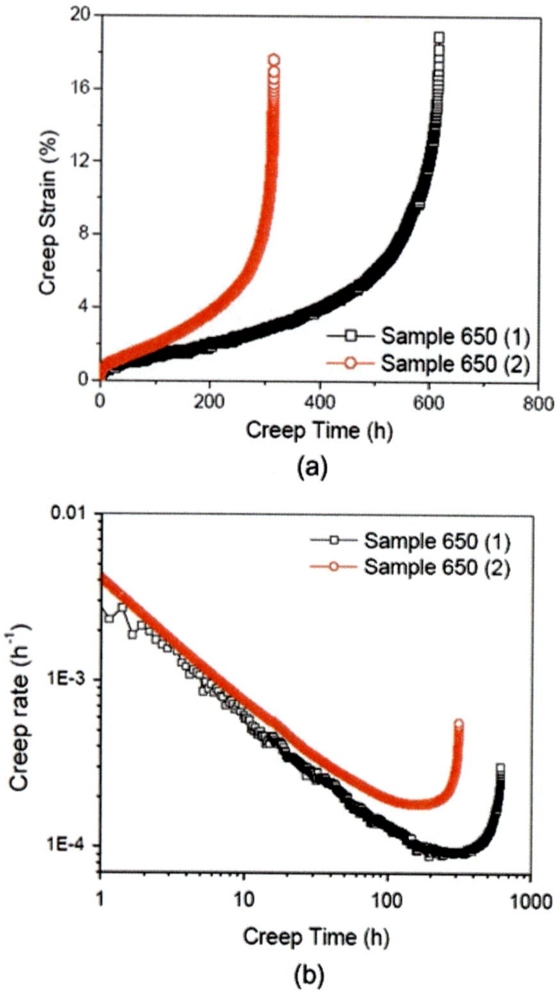

Fig. 3.3 Creep properties of P91 at 650 °C under 100 MPa stress. **a** Creep strain and **b** Creep strain rate, as a function of time

Chapter 4
Micrograph Description

4.1 As-Received Pipe

Figures 4.1, 4.2, 4.3, 4.4 and 4.5.

4.2 Micrographs of Creep Interrupted P91: 550 °C Temperature and 200 MPa Stress

Figures 4.6, 4.7, 4.8, 4.9, 4.10, 4.11, 4.12, 4.13, 4.14, 4.15, 4.16, 4.17, 4.18, 4.19, 4.20, 4.21, 4.22, 4.23, 4.24, 4.25, 4.26, 4.27, 4.28, 4.29, 4.30, 4.31, 4.32, 4.33, 4.34, 4.35, 4.36 and 4.37.

4.3 Micrographs of Creep Interrupted P91: 600 °C Temperature and 140 MPa Stress

Figures 4.38, 4.39, 4.40, 4.41, 4.42, 4.43, 4.44, 4.45, 4.46, 4.47, 4.48, 4.49, 4.50, 4.51, 4.52, 4.53, 4.54, 4.55, 4.56, 4.57, 4.58, 4.59, 4.60, 4.61, 4.62 and 4.63.

4.4 Micrographs of Creep Interrupted P91: 650 °C Temperature and 100 MPa Stress

Figures 4.64, 4.65, 4.66, 4.67, 4.68, 4.69, 4.70, 4.71, 4.72, 4.73, 4.74, 4.75, 4.76, 4.77, 4.78, 4.79, 4.80, 4.81, 4.82, 4.83, 4.84 and 4.85.

R. K. Roy et al., *Microstructure Atlas of P91 Steel*,
https://doi.org/10.1007/978-981-19-9562-0_4

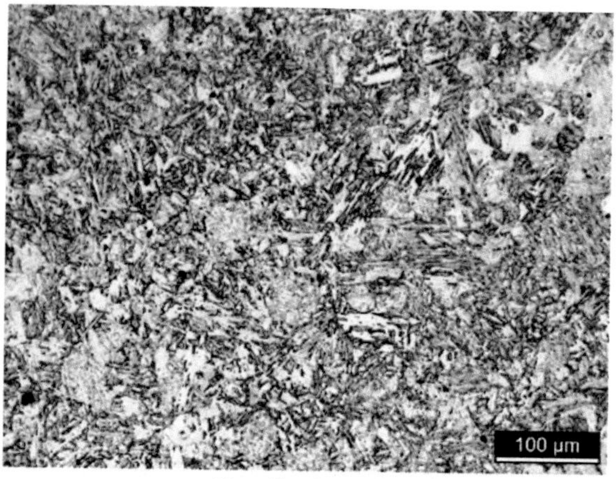

Magnification 200X

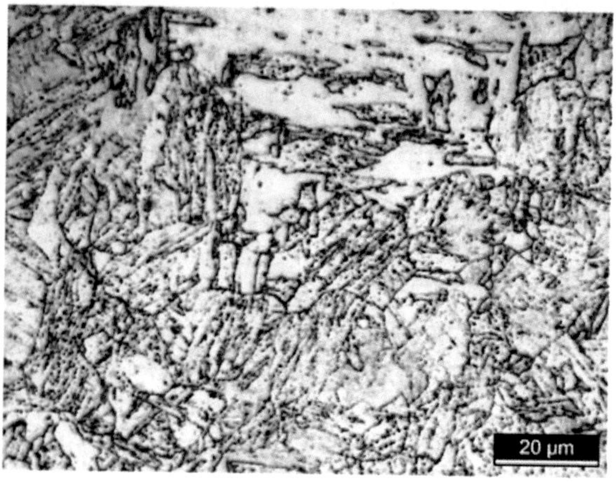

Magnification 1000X

Fig. 4.1 Optical micrographs of as-received P91 (9Cr-1Mo) steel at transverse section showing martensite structure with prior austenitic grain boundary

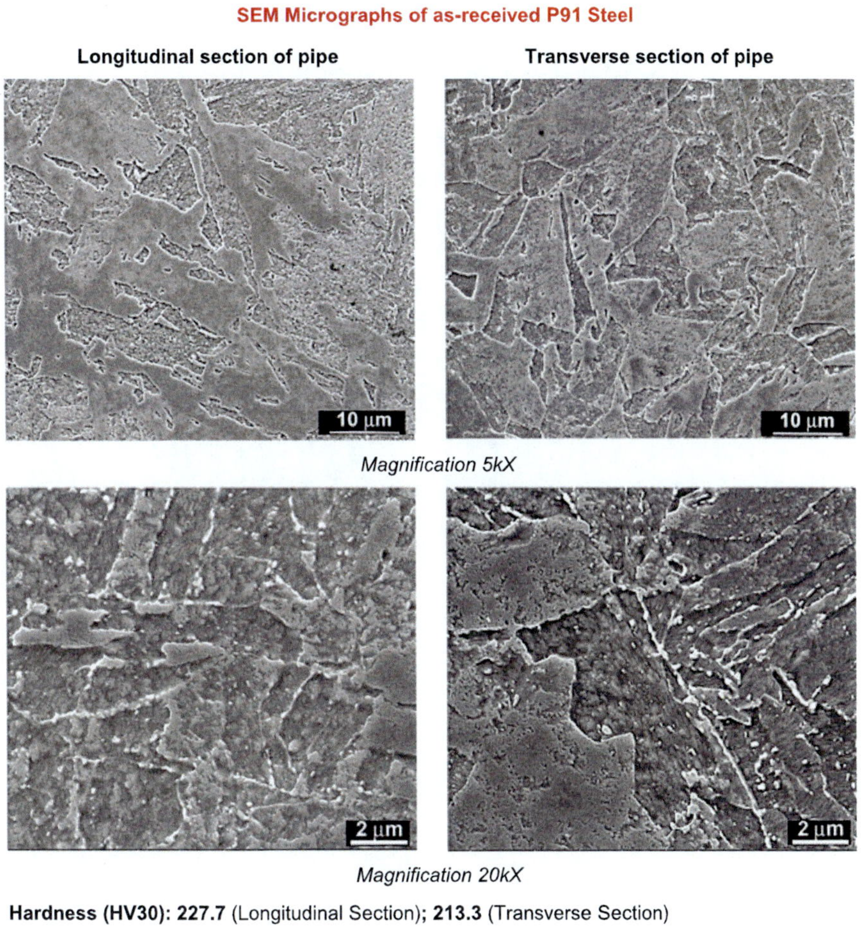

Fig. 4.2 SEM-SE micrographs of as-received P91 (9Cr-1Mo) steel, showing distribution of fine carbide precipitates at grain boundary and grain interior of lath martensite

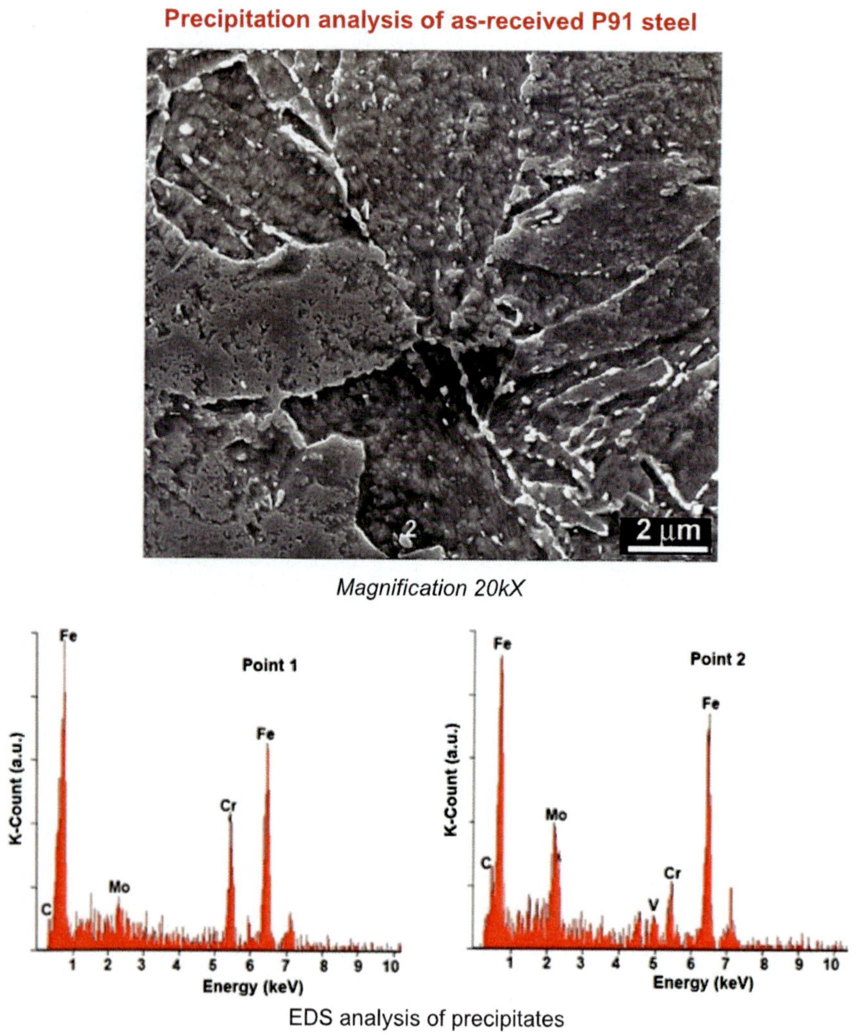

Fig. 4.3 SEM-SE micrographs (*magnification 20kX)* of as-received P91 (9Cr-1Mo) steel showing distribution of fine Cr-rich (point 1) $M_{23}C_6$ (M = Cr, Mo Fe) carbide precipitates and Mo-rich (point 2) phases. It is noteworthy that Cr-rich precipitates predominantly occur along grain boundary, whereas Mo-rich precipitates occur mostly within grain interior

TEM Micrographs of as-received P91 steel

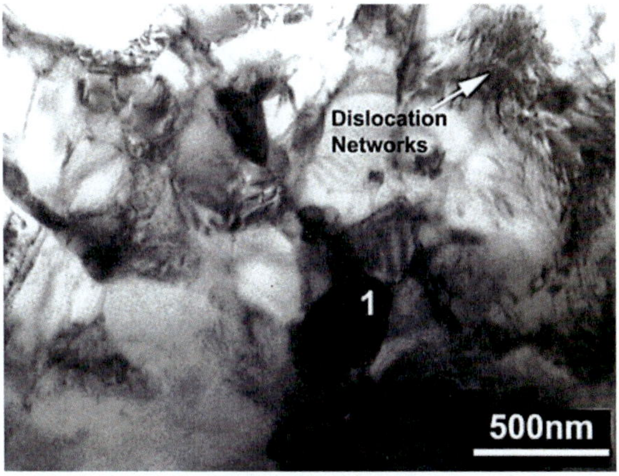

Magnification 11.5kX

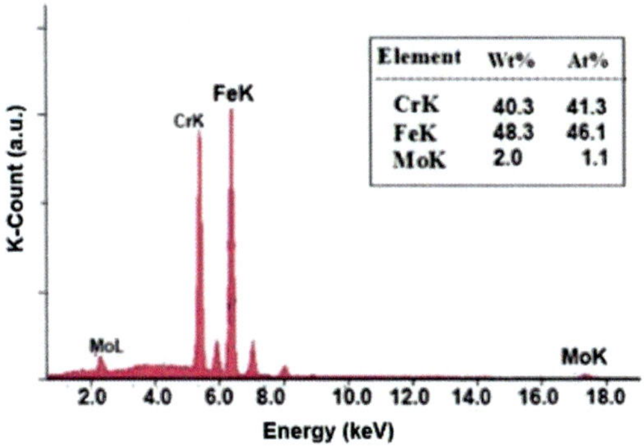

EDS analysis of precipitate (point 1)

Fig. 4.4 TEM-BF image of specimen from ***transverse section*** of as-received P91 steel, showing irregular-shaped precipitates distributed at grain boundary. Jumbled dislocations were present at the close vicinity of precipitate. EDS elemental analysis (point1) illustrates that the precipitate was Cr-rich carbide with small amount of Mo

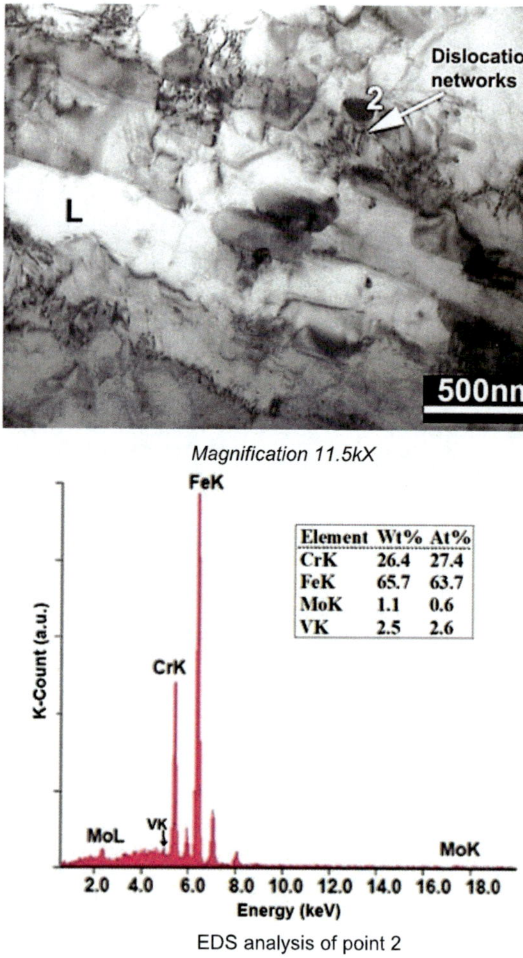

Element	Wt%	At%
CrK	26.4	27.4
FeK	65.7	63.7
MoK	1.1	0.6
VK	2.5	2.6

Fig. 4.5 TEM-BF image from ***transverse section*** of as-received P91 steel, showing hindrance of dislocation motion against irregular-shaped second phase. Lack of precipitation at localized region propels smooth dislocation gliding to reveal relatively clean matrix (L). EDS spectrum contains Cr–Mo characteristic peaks with a small peak of V

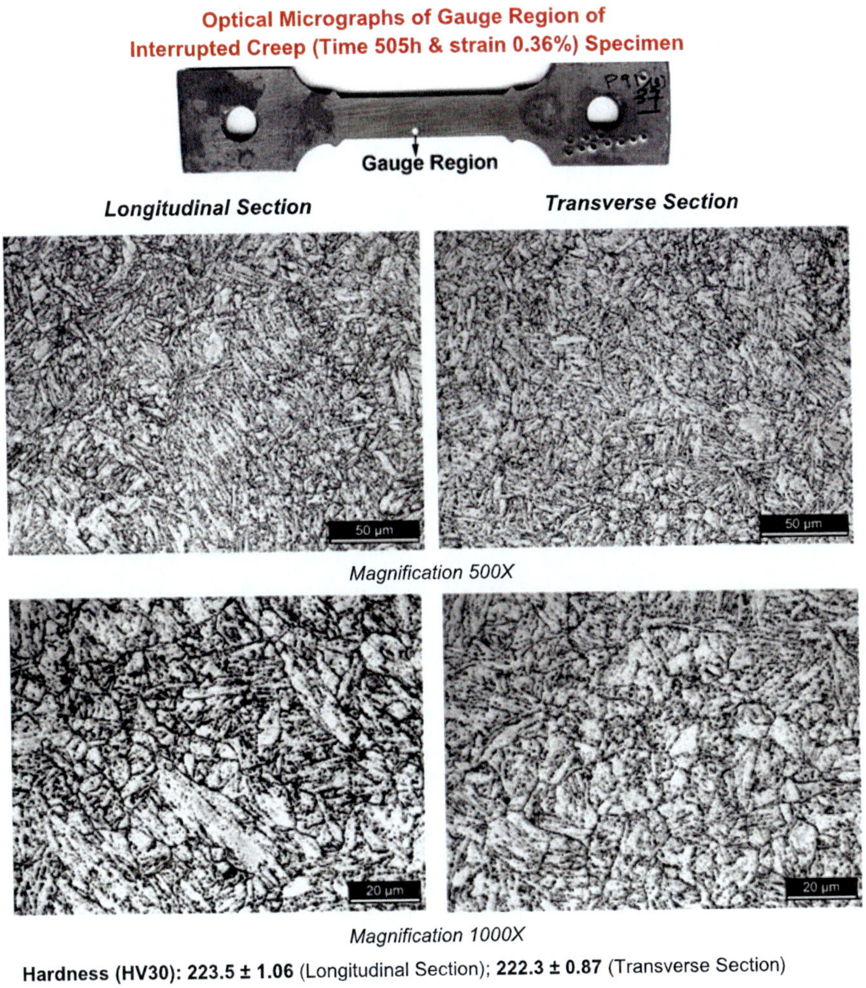

Fig. 4.6 Optical micrographs of **longitudinal and transverse sections** of P91 (9Cr-1Mo) steel after creep exposed at 550 °C/200 MPa for 505 h, showing martensitic structures

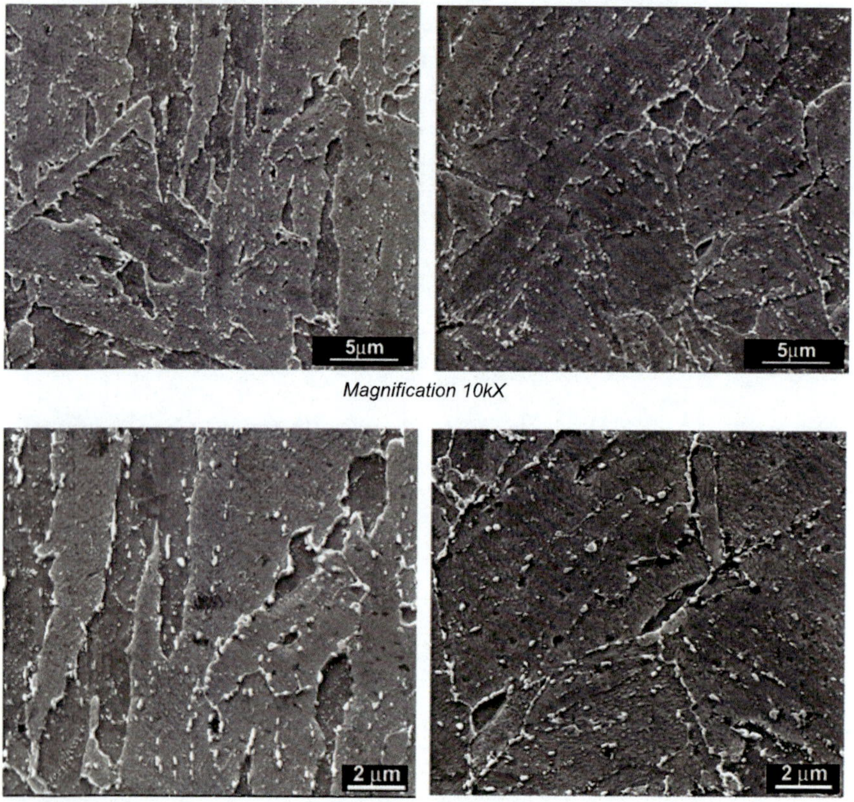

Fig. 4.7 SEM-SE micrographs of **longitudinal and transverse sections** of P91 (9Cr-1Mo) steel after creep exposed at 550 °C/200 MPa for 505 h, showing the distribution of carbides at grain interiors of martensitic structures

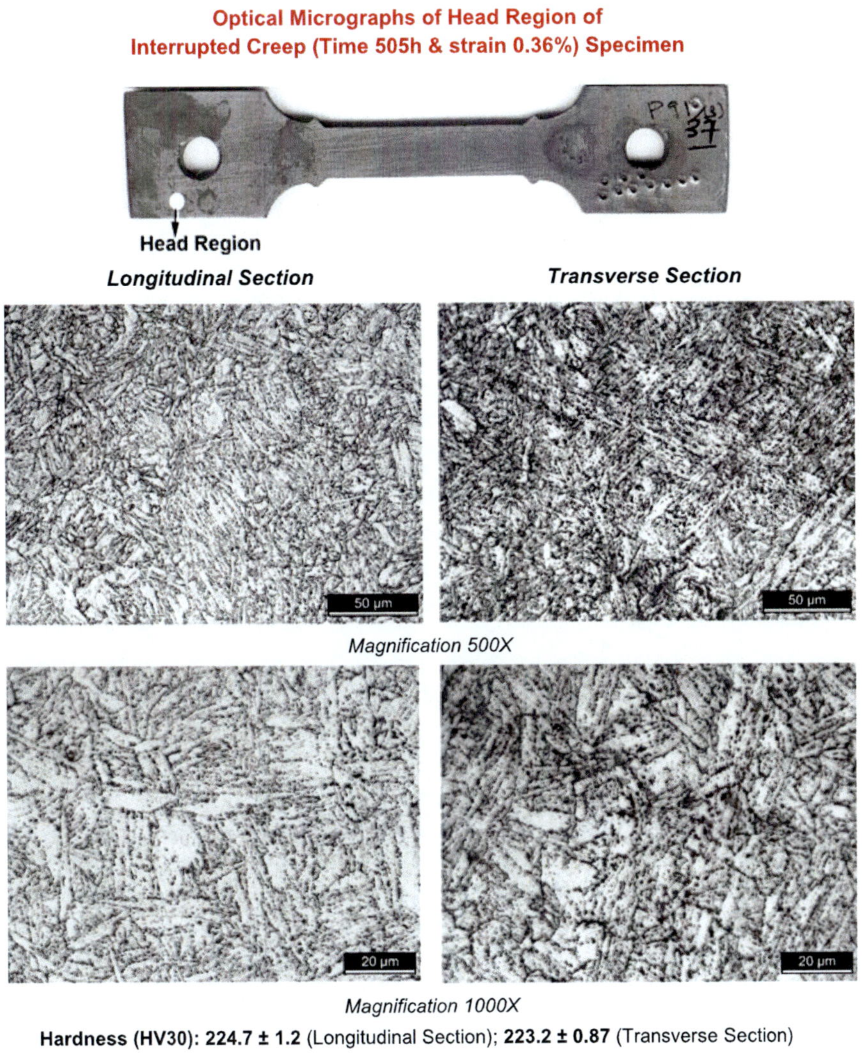

Fig. 4.8 Optical micrographs of **longitudinal and transverse sections** of P91 (9Cr-1Mo) steel after creep exposed at 550 °C/200 MPa for 505 h, showing martensitic structures

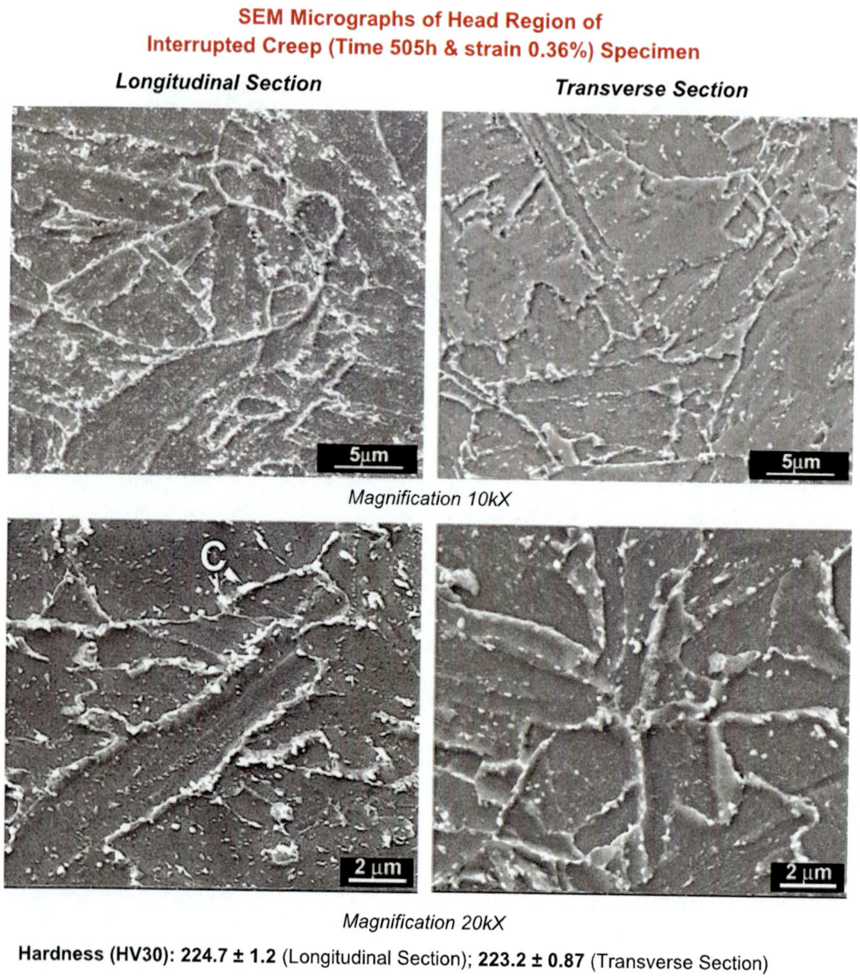

Fig. 4.9 SEM-SE micrographs of **longitudinal and transverse sections** of P91 (9Cr-1Mo) steel after creep exposed at 550 °C/200 MPa for 505 h, showing the distribution of carbides (C) at grain interiors of martensitic matrix

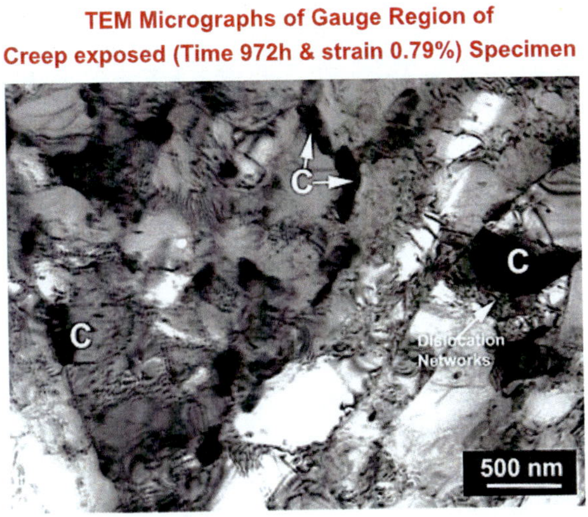

Fig. 4.10 TEM-BF image of <u>*gauge region*</u> at ***transverse section*** of P91(9Cr-1Mo) steel after creep exposed at 550 °C/200 MPa for 972 h with a creep strain of 0.79%, explaining $M_{23}C_6$ carbides (C) precipitated at grain boundary and grain interiors, interacting to dislocation networks. Signature of localized recrystallization was there by the presence of polygonal grain clustering

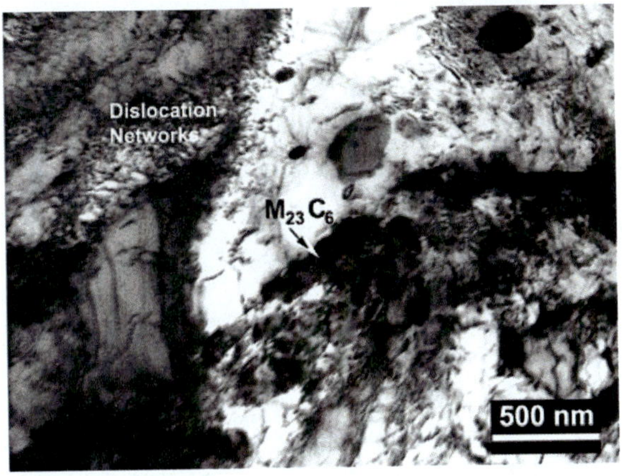

Fig. 4.11 TEM-BF image of <u>*gauge region*</u> at ***transverse section*** of P91(9Cr-1Mo) steel after creep exposed at 550 °C/200 MPa for 972 h with a creep strain of 0.79%, explaining $M_{23}C_6$ carbides precipitated at grain boundary and grain interiors and dislocation networks throughout the matrix

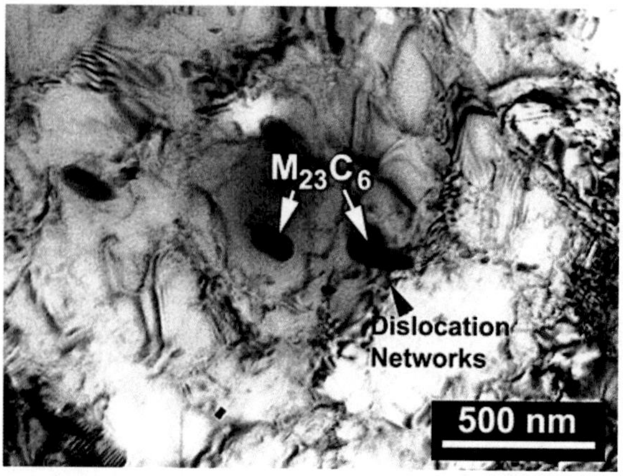

Fig. 4.12 TEM-BF image of <u>*gauge region*</u> at ***transverse section*** of P91(9Cr-1Mo) steel after creep exposed at 550 °C/200 MPa for 972 h with a creep strain of 0.79%, explaining $M_{23}C_6$ carbides precipitated at grain boundary and grain interiors, interacting to dislocations at many places

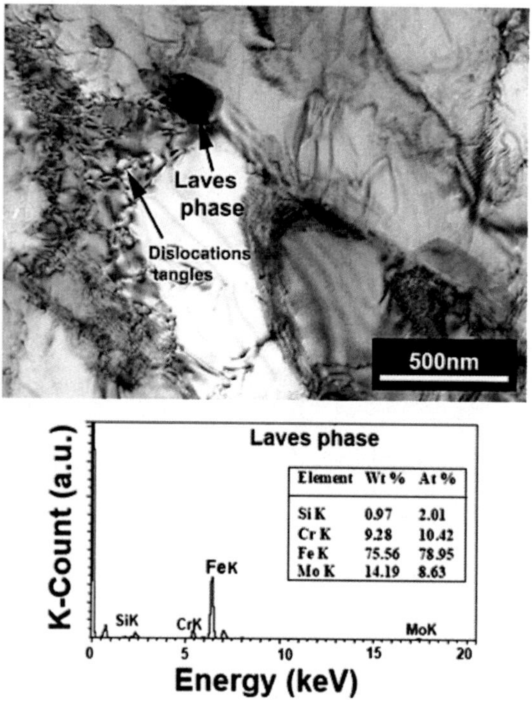

EDS analysis of Laves phase

Fig. 4.13 TEM-BF image of <u>*gauge region*</u> at ***transverse section*** of P91(9Cr-1Mo) steel after creep exposed at 550 °C/200 MPa for 972 h with a creep strain of 0.79%, explaining Laves phase precipitated at grain boundary and interaction with dislocations formed as tangles. EDS analysis confirms the Mo-rich Laves phase, consisting of small amount of Si

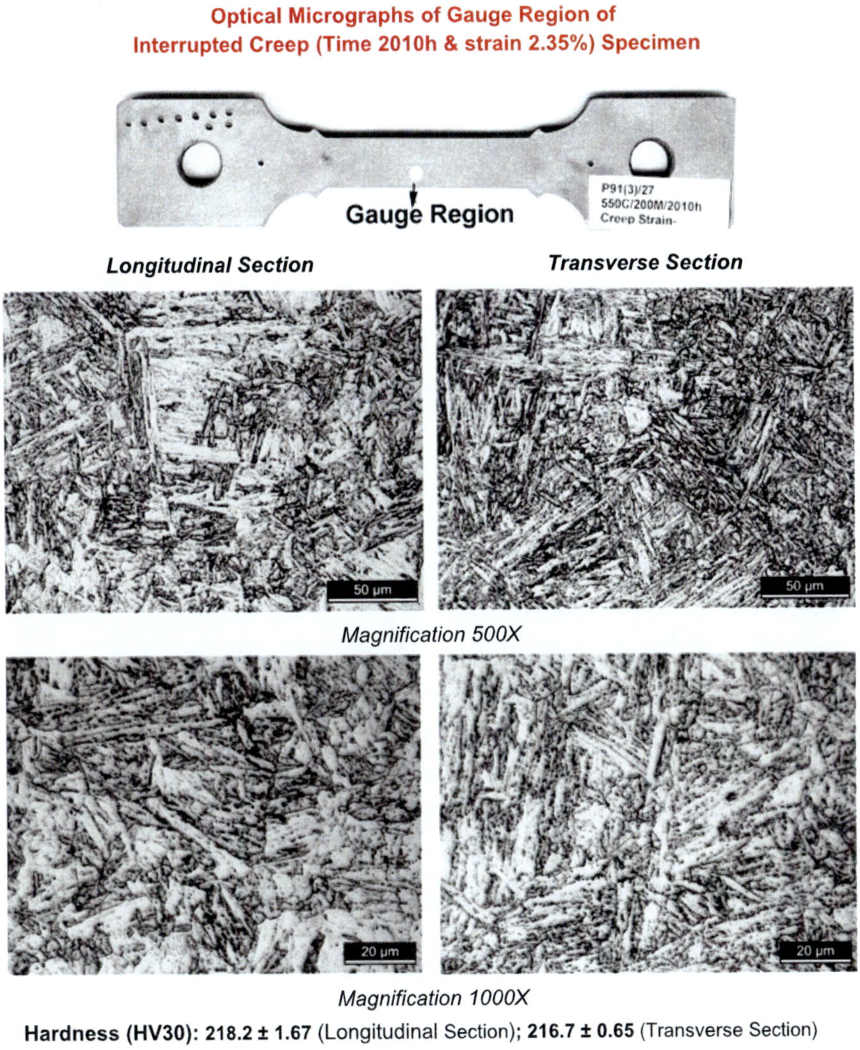

Fig. 4.14 Optical micrographs of **longitudinal and transverse sections** of P91 (9Cr-1Mo) steel after creep exposed at 550 °C/200 MPa for 2010 h, showing martensitic structures

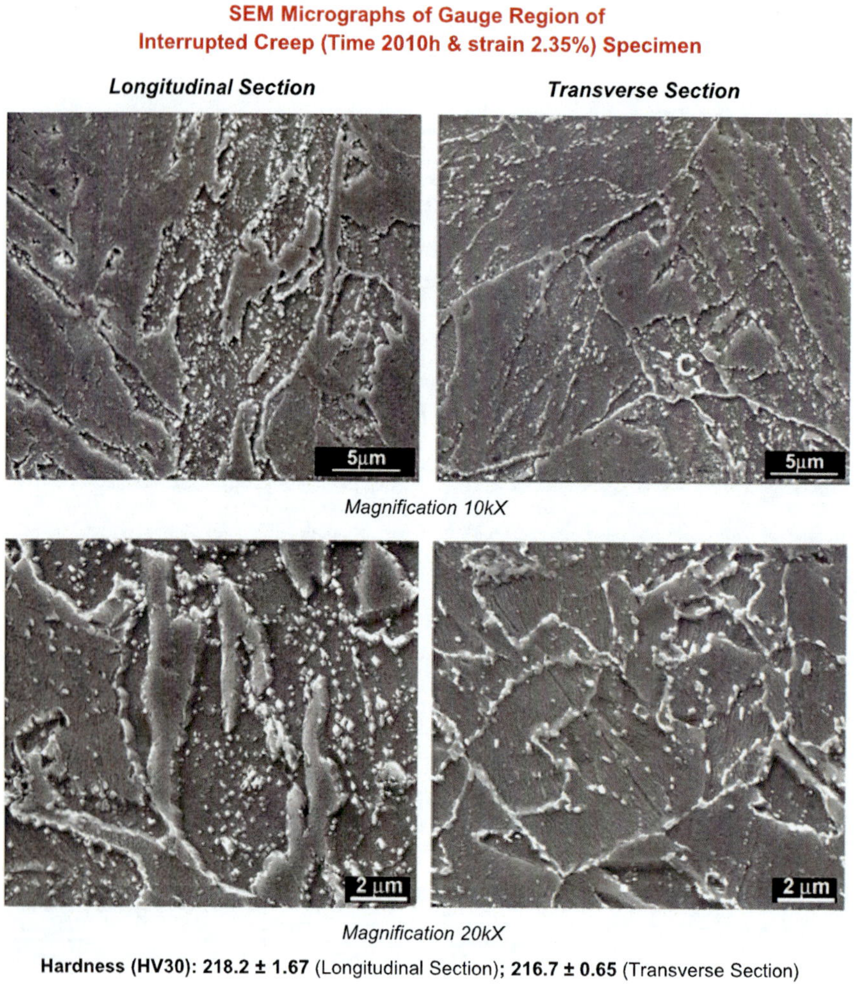

Fig. 4.15 SEM-SE micrographs of **longitudinal and transverse sections** of P91 (9Cr-1Mo) steel after creep exposed at 550 °C/200 MPa for 2010 h, showing carbide (C) precipitates at grain interiors and grain boundaries

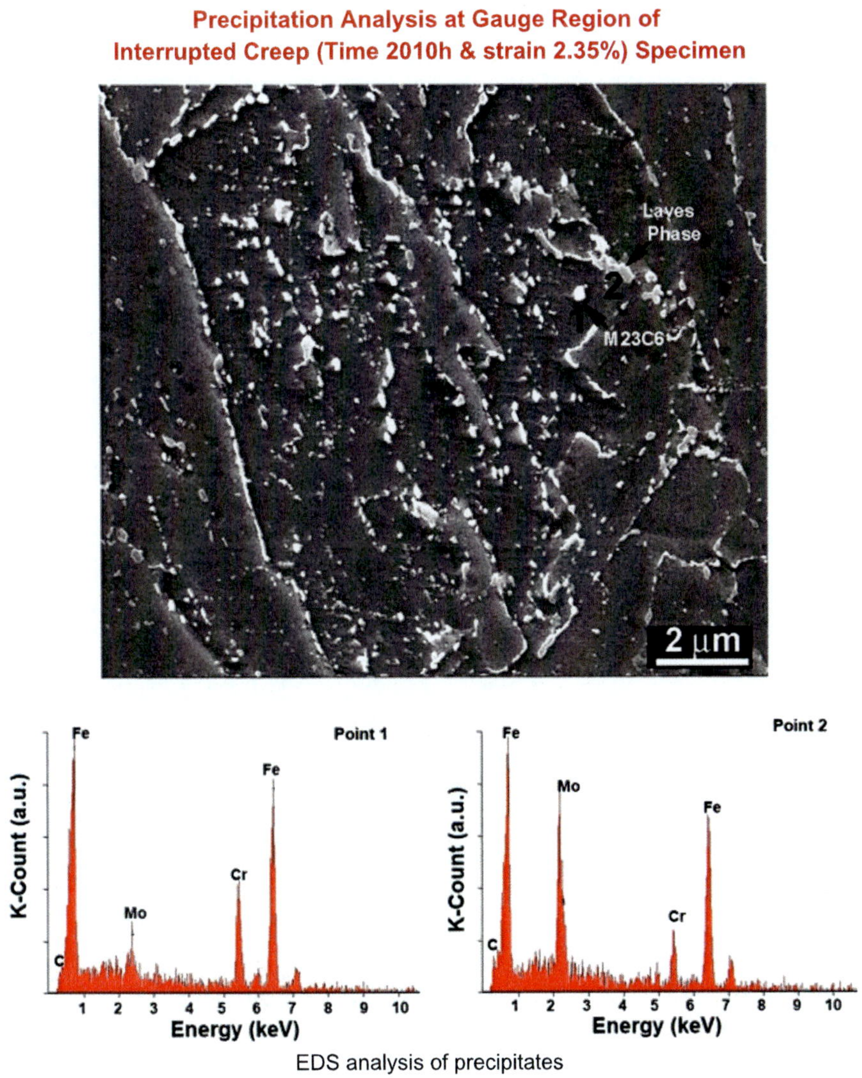

Fig. 4.16 SEM-SE micrograph (magnification 20kX) of ***transverse sections*** of P91 (9Cr-1Mo) steel after creep exposed at 550 °C/200 MPa for 2010 h, showing Cr-rich $M_{23}C_6$ [M = Fe, Cr, Mo] (point 1) and Mo-rich Laves phase (point 2) precipitates at grain interiors and grain boundaries of martensite matrix

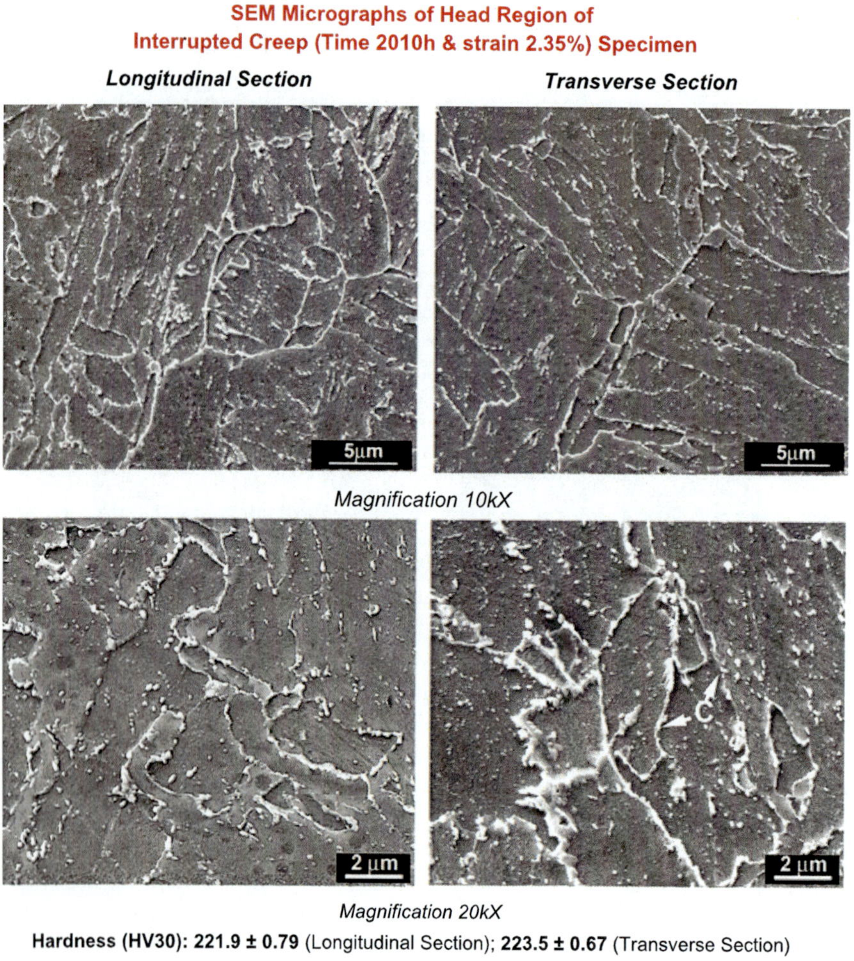

Fig. 4.17 SEM-SE micrographs of **longitudinal and transverse sections** of P91 (9Cr-1Mo) steel after creep exposed at 550 °C/200 MPa for 2010 h, showing Cr-rich $M_{23}C_6$ (C) and Mo-rich Laves phase precipitates at grain interiors and grain boundaries of martensite matrix

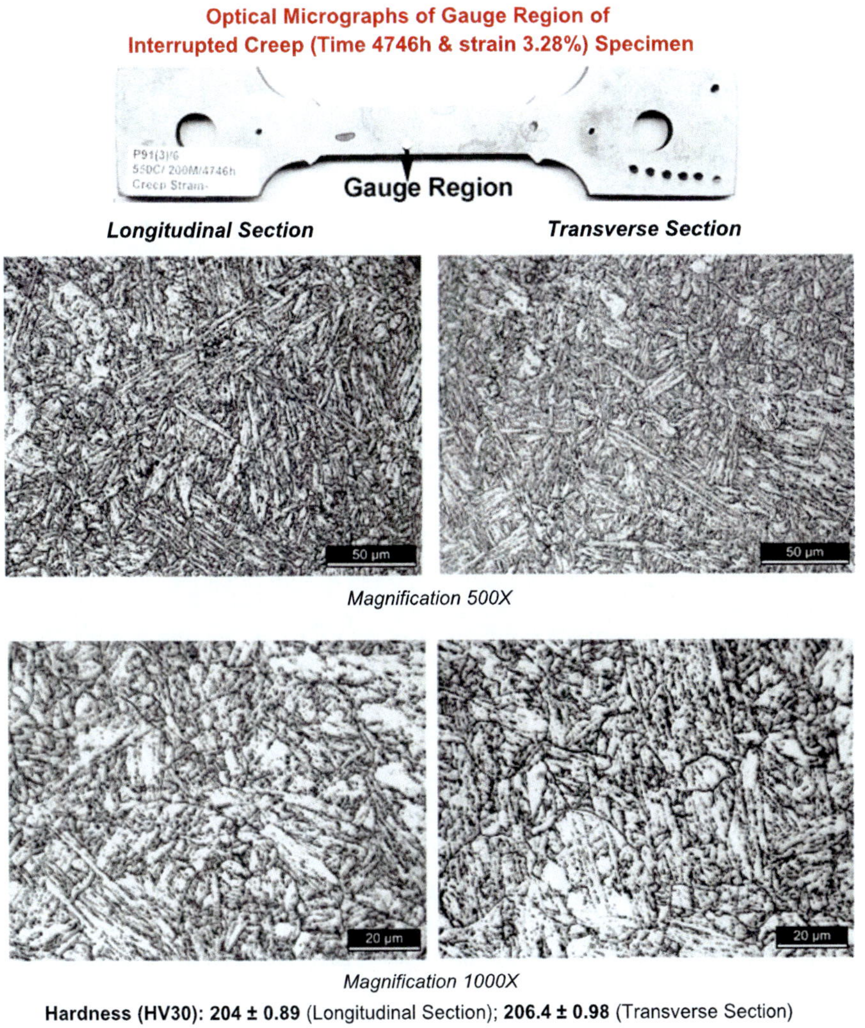

Fig. 4.18 Optical micrographs of **longitudinal and transverse sections** of P91 (9Cr-1Mo) steel after creep exposed at 550 °C/200 MPa for 4746 h, showing martensitic structures

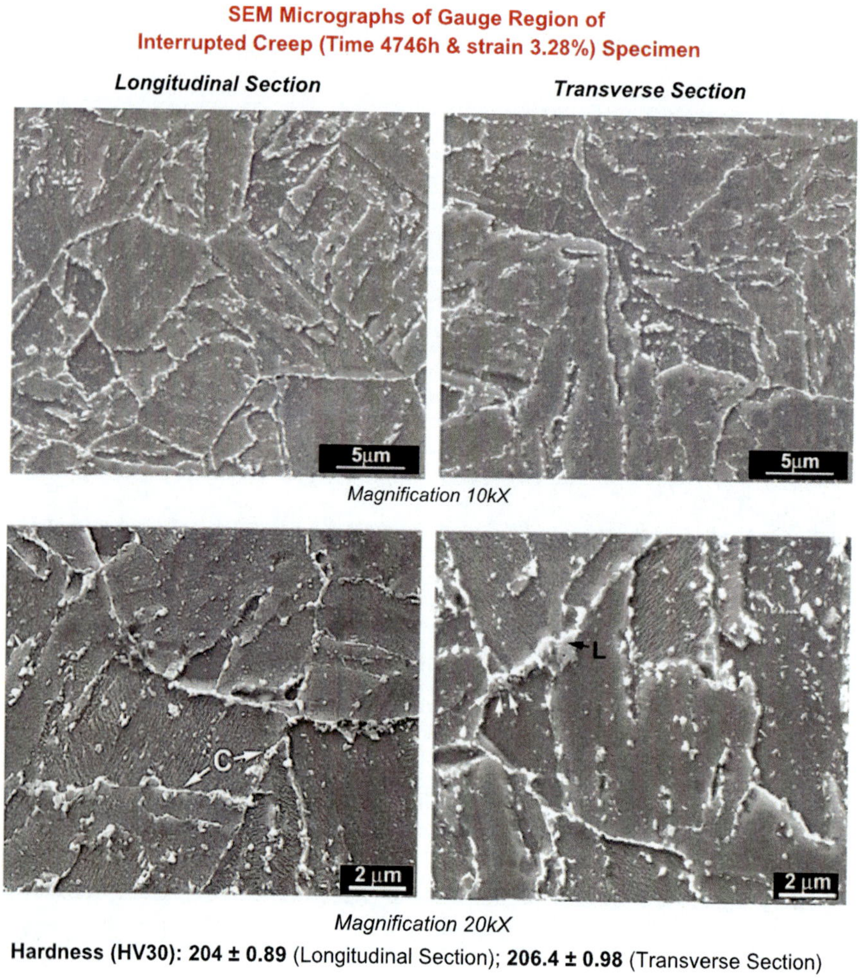

Fig. 4.19 SEM-SE micrographs of **longitudinal and transverse sections** of P91 (9Cr-1Mo) steel after creep exposed at 550 °C/200 MPa for 4746 h, Cr-rich $M_{23}C_6$ (C) and Mo-rich Laves phase (L) precipitates at grain interiors and grain boundaries of martensitic structures. White arrows indicating coalescence of precipitates at grain boundary

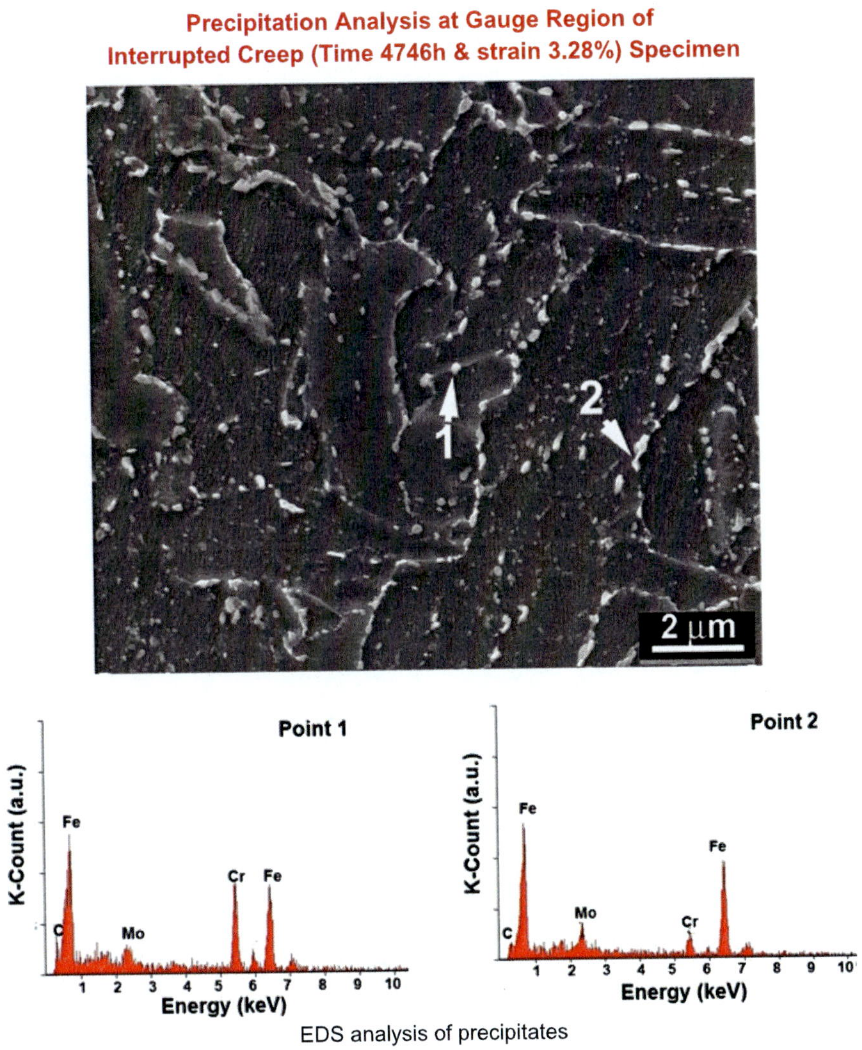

Fig. 4.20 SEM-SE micrograph (magnification 20kX) of ***transverse sections*** of P91 (9Cr-1Mo) steel after creep exposed at 550 °C/200 MPa for 4746 h, showing Cr-rich $M_{23}C_6$ [M = Fe, Cr, Mo] (point 1) and Mo-rich Laves phase (point 2) precipitates at grain interiors and grain boundaries of martensite matrix

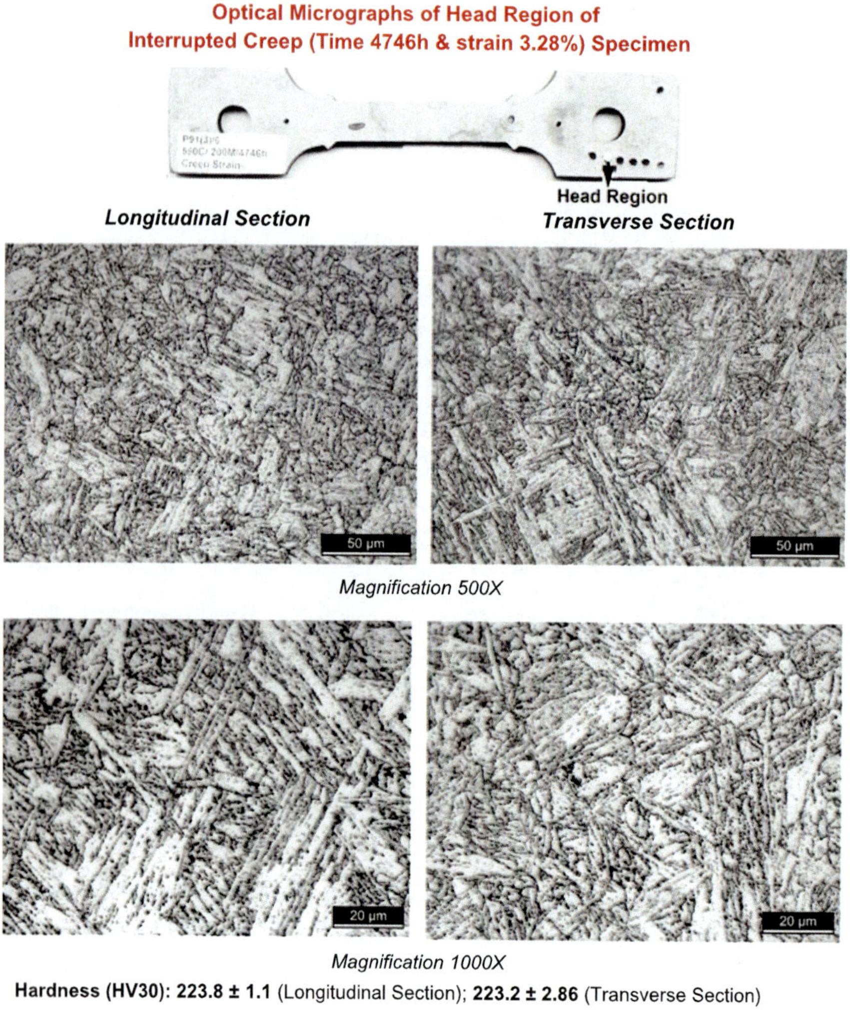

Fig. 4.21 Optical micrographs of **longitudinal and transverse sections** of P91 (9Cr-1Mo) steel after creep exposed at 550 °C/200 MPa for 4746 h, showing martensitic structures

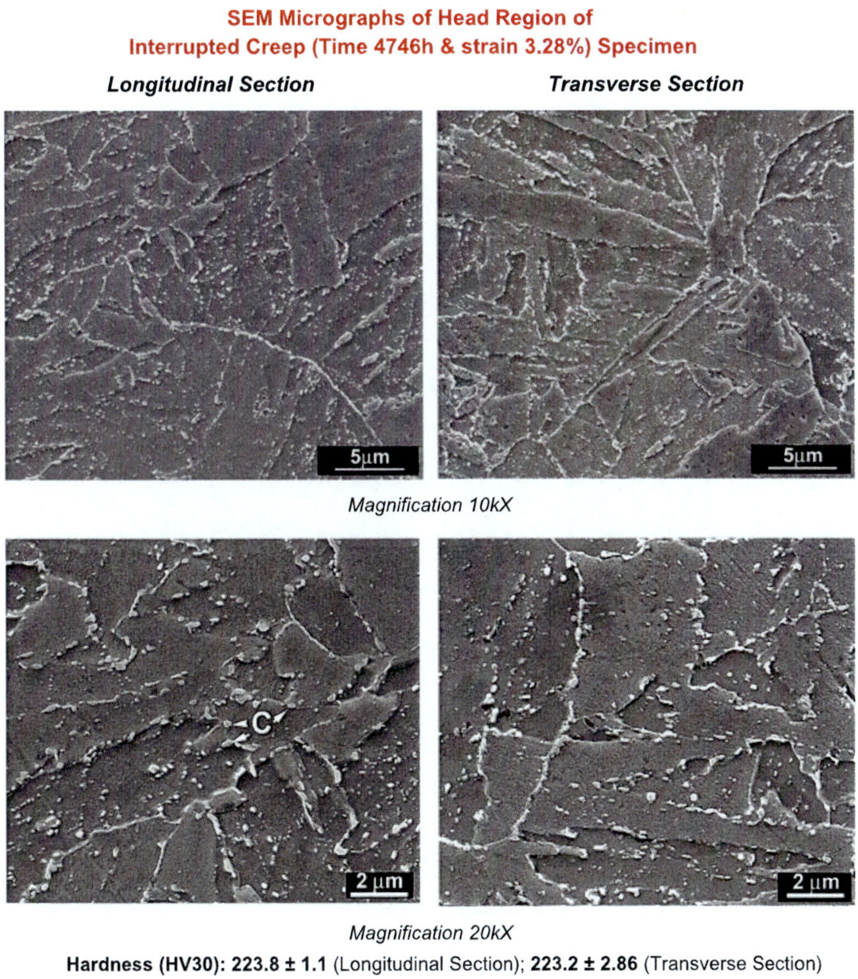

Fig. 4.22 SEM-SE micrographs of **longitudinal and transverse sections** of P91 (9Cr-1Mo) steel after creep exposed at 550 °C/200 MPa for 4746 h, showing carbides (C) at grain interiors and grain boundaries of martensite matrix

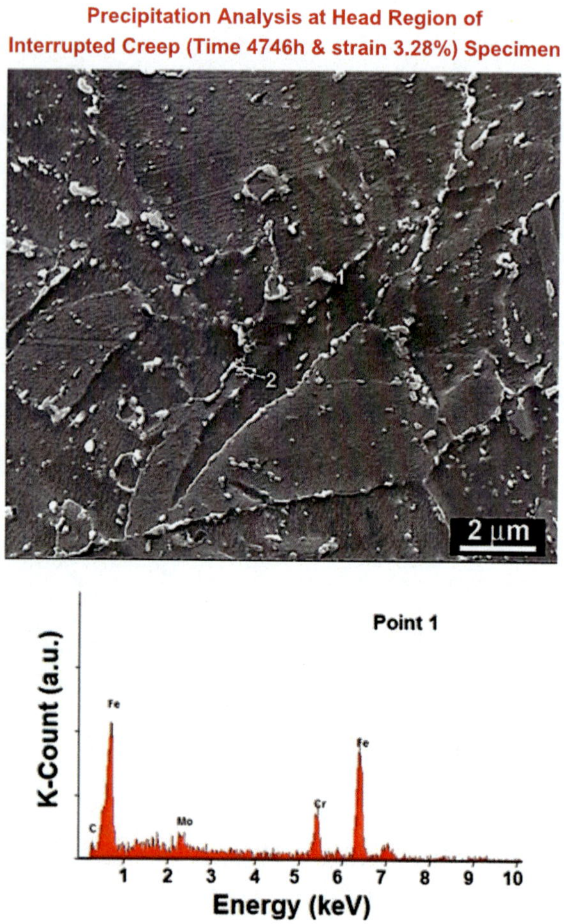

EDS analysis of precipitate (point 1)

Fig. 4.23 SEM-SE micrograph (magnification 20kX) of ***longitudinal section*** of P91 (9Cr-1Mo) steel after creep exposed at 550 °C/200 MPa for 4746 h, showing Cr-rich $M_{23}C_6$ (point 1 and point 2) precipitates at grain interiors and grain boundaries of martensite matrix

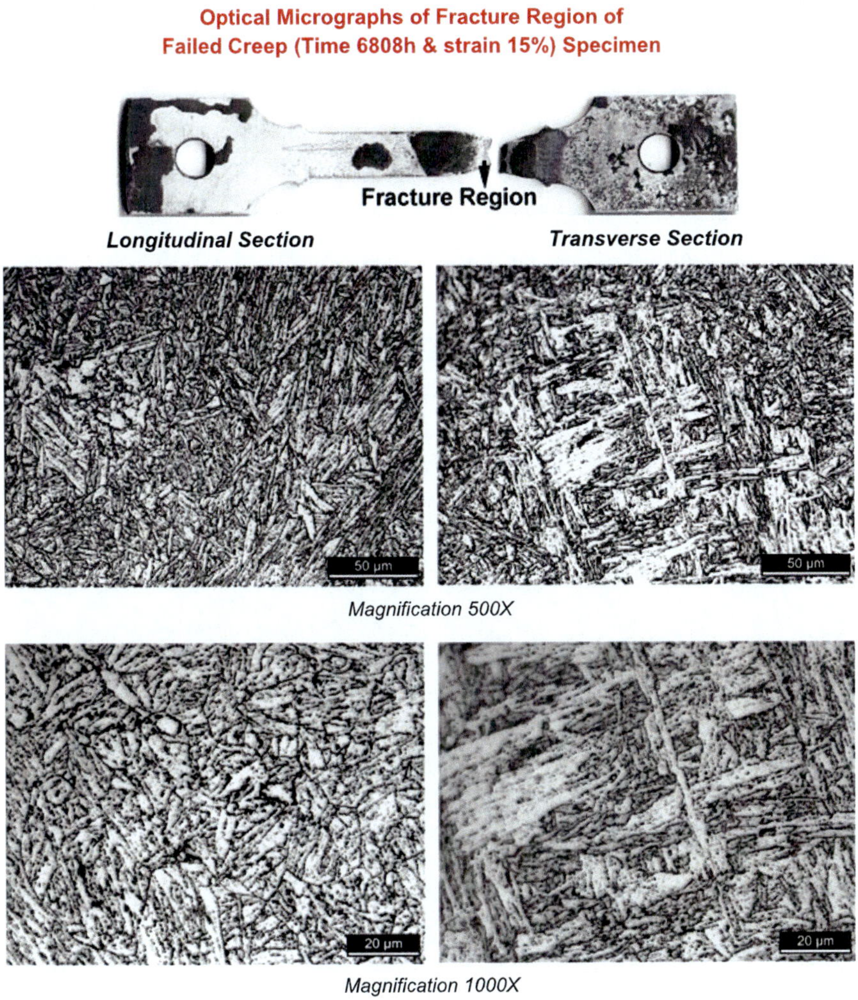

Fig. 4.24 Optical micrographs of **longitudinal and transverse sections** of P91 (9Cr-1Mo) steel after creep exposed at 550 °C/200 MPa for 6808 h, showing martensitic structures

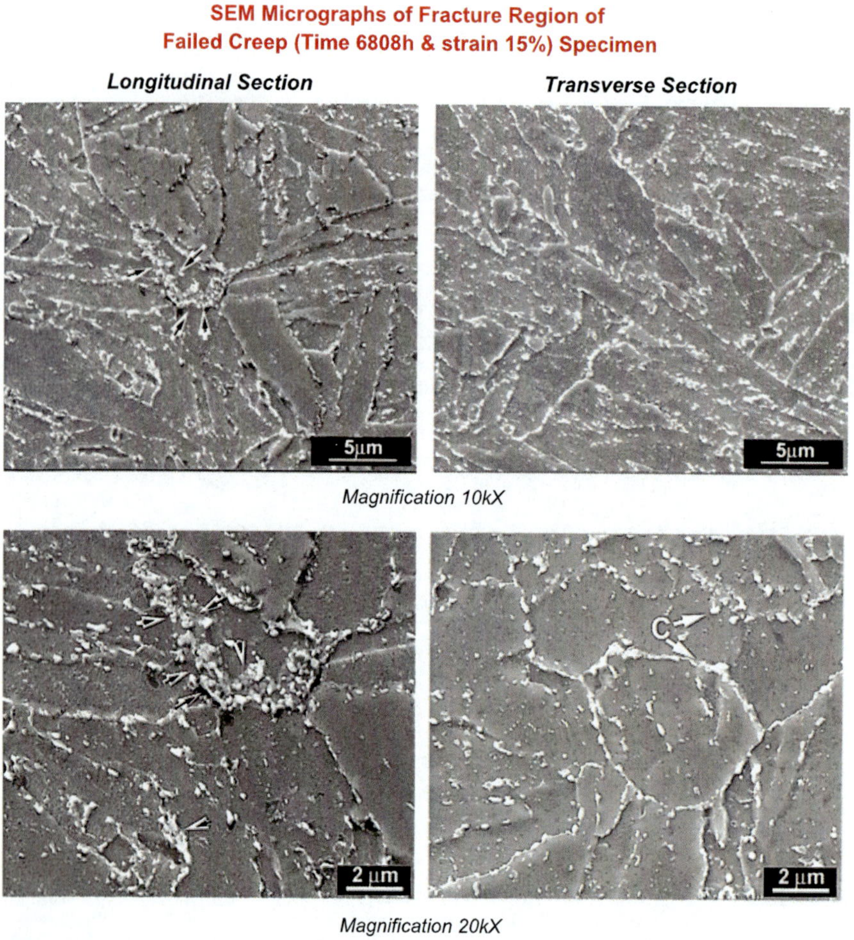

Fig. 4.25 SEM-SE micrographs of **longitudinal and transverse sections** of P91 (9Cr-1Mo) steel after creep exposed at 550 °C/200 MPa for 6808 h, showing inhomogeneous distribution of carbide (C) precipitates at grain boundary and grain interiors with the accumulation of large amounts carbides at some point of grain boundaries (black-white arrows)

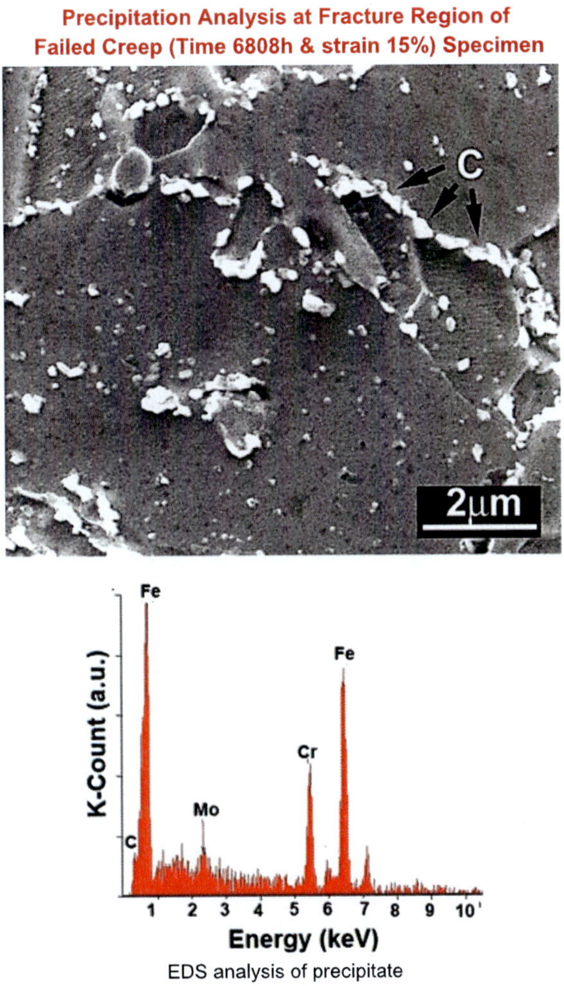

Fig. 4.26 SEM-SE micrograph (magnification 30kX) of ***transverse section*** of P91 (9Cr-1Mo) steel after creep exposed at 550 °C/200 MPa for 6808 h, showing Cr-rich $M_{23}C_6$ [M = Fe, Cr, Mo] (point C) precipitates at grain interiors and grain boundaries of martensitic structures. Some carbides are distributed like a chain at grain boundaries (showing by black arrows)

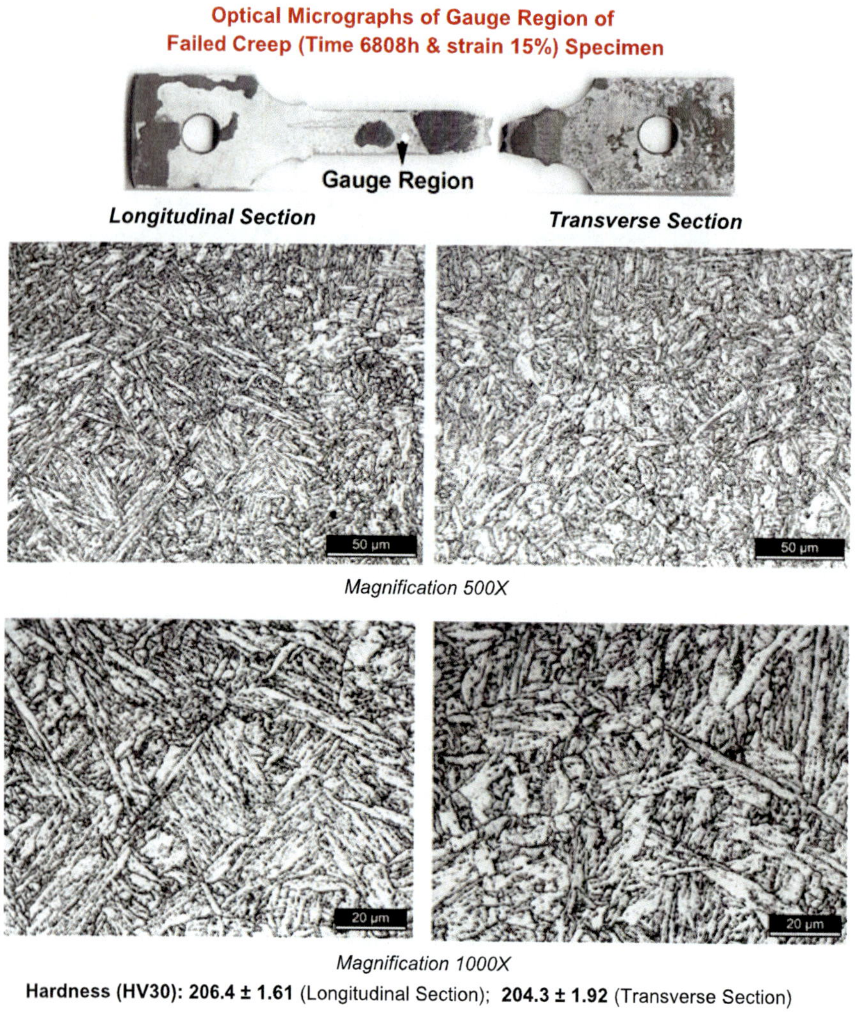

Fig. 4.27 Optical micrographs of **longitudinal and transverse sections** of P91 (9Cr-1Mo) steel after creep exposed at 550 °C/200 MPa for 6808 h, explaining martensitic structures

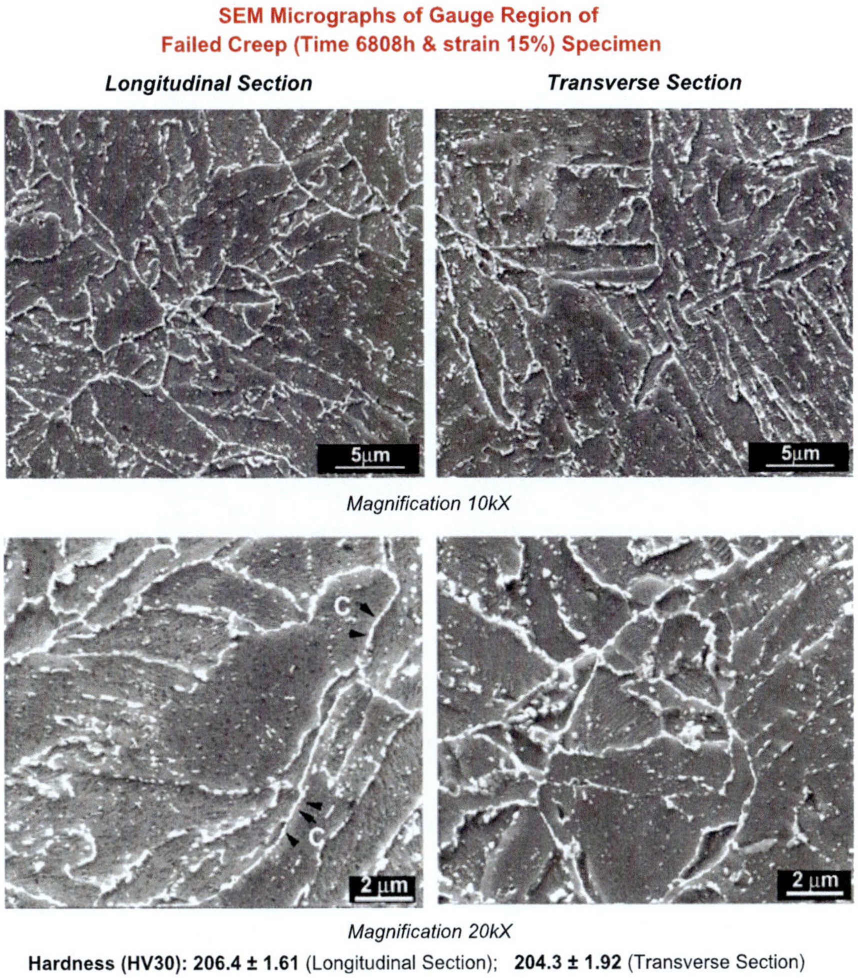

Fig. 4.28 SEM-SE micrographs of **longitudinal and transverse sections** of P91 (9Cr-1Mo) steel after creep exposed at 550 °C/200 MPa for 6808 h, showing martensitic structures

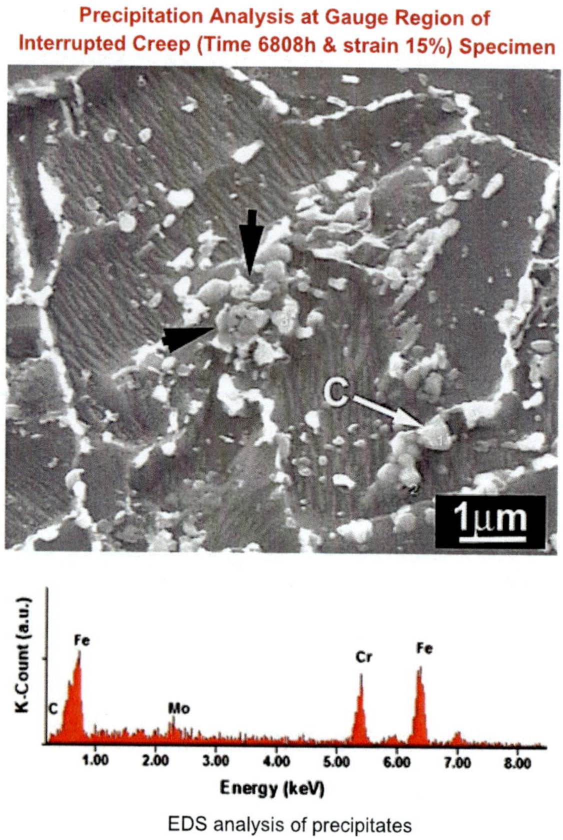

Fig. 4.29 SEM-SE micrograph (magnification 40kX) of ***transverse sections*** of P91 (9Cr-1Mo) steel after creep exposed at 550 °C/200 MPa for 6808 h, showing Cr-rich $M_{23}C_6$ [M = Fe, Cr, Mo] (point C) precipitates at grain interiors and grain boundaries of martensitic structures. Some carbide agglomerated at center regions of grain (shown by black arrows)

TEM Micrographs of Gauge Region of
Failed Creep (Time 6808h & strain 15%) Specimen

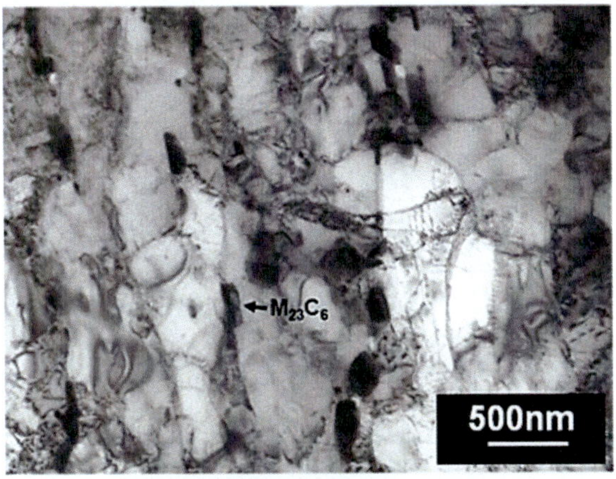

Magnification 6.6kX

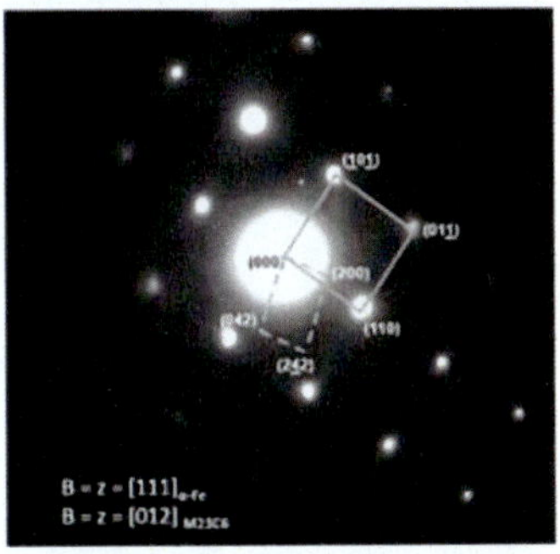

SAED pattern of carbide precipitate

Fig. 4.30 Transmission electron micrograph bright field image of <u>*gauge region*</u> at ***transverse section*** of P91(9Cr-1Mo) steel after creep failed at 550 °C/200 MPa for 6808 h with a creep strain of 15%, showing $M_{23}C_6$ carbides (with SAED pattern) precipitated at grain boundary. Traces of lath structure are still prevalent

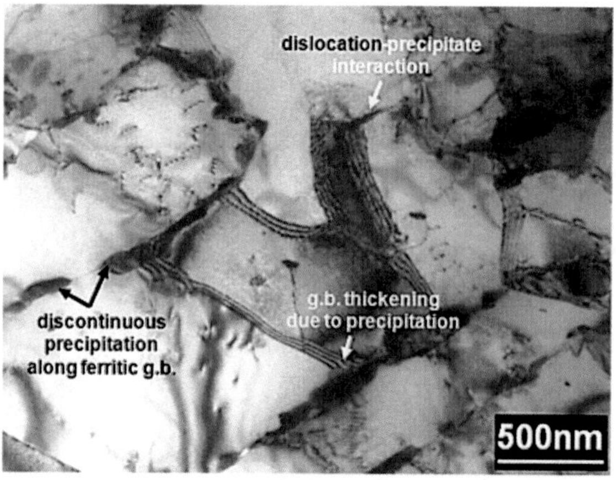

Magnification 8.8kX

Fig. 4.31 Transmission electron micrograph bright field image of gauge region at ***transverse section*** of P91(9Cr-1Mo) steel after creep failed at 550 °C/200 MPa for 6808 h with a creep strain of 15%, showing $M_{23}C_6$ carbides (white arrows) precipitated at grain boundary and interacting with dislocation lines (black arrows). Thickening of grain boundary is the indication discontinues network of precipitates and formation of dislocation trap

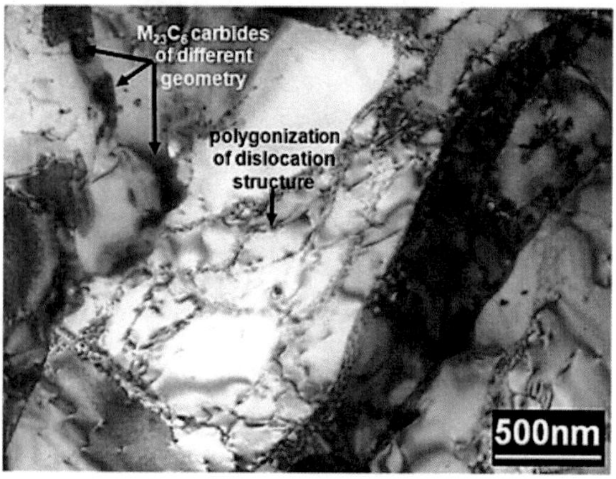

Magnification 8.8kX

Fig. 4.32 Transmission electron micrograph bright field image of gauge region at ***transverse section*** of P91(9Cr-1Mo) steel after creep failed at 550 °C/200 MPa for 6808 h with a creep strain of 15%, explaining the distribution of $M_{23}C_6$ carbides of different geometry and the polygonization of dislocation at matrix

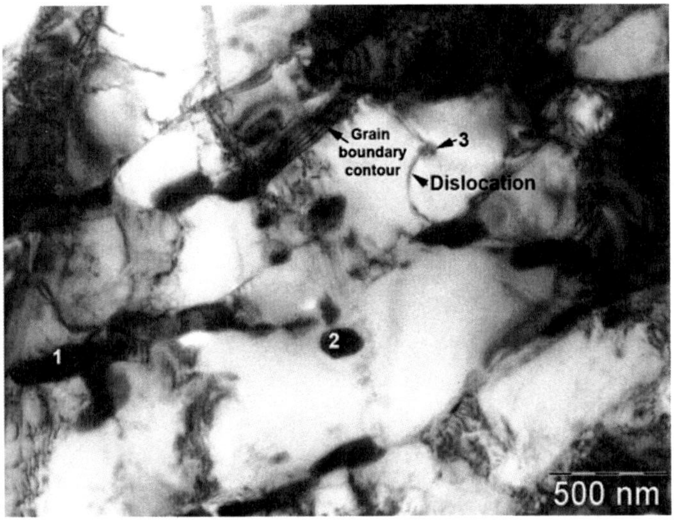

Magnification 8.8kX

Fig. 4.33 Transmission electron micrograph bright field image of gauge region at ***transverse section*** of P91(9Cr-1Mo) steel after creep failed at 550 °C/200 MPa for 6808 h with a creep strain of 15%, explaining $M_{23}C_6$ carbides of varying shaped parallelepiped (point 1) at grain boundary, oval (point 2), and spherical (point 3) while dislocation lines and dislocation interacting to particles (point 3)

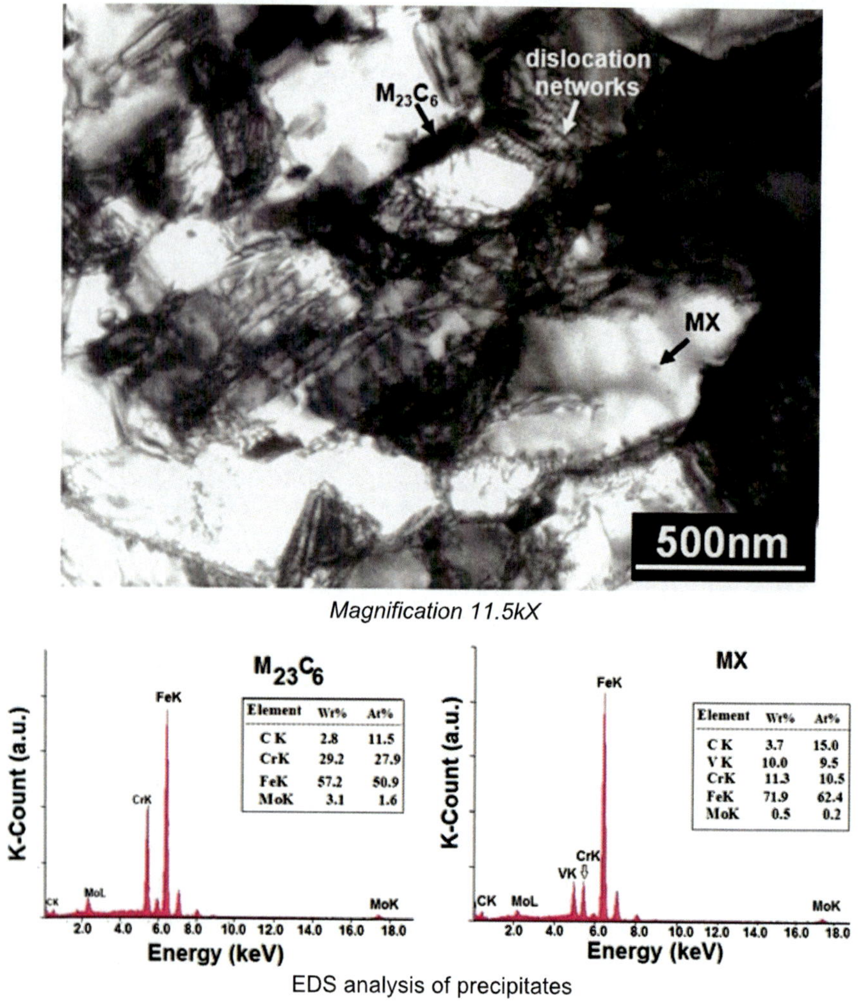

Element	Wt%	At%
C K	2.8	11.5
CrK	29.2	27.9
FeK	57.2	50.9
MoK	3.1	1.6

Element	Wt%	At%
C K	3.7	15.0
V K	10.0	9.5
CrK	11.3	10.5
FeK	71.9	62.4
MoK	0.5	0.2

Fig. 4.34 TEM-BF image of <u>*gauge region*</u> at ***transverse section*** of P91(9Cr-1Mo) steel after creep failed at 550 °C/200 MPa for 6808 h with a creep strain of 15%, showing rectangular $M_{23}C_6$ carbides at grain boundary, and interacting to dislocation networks. Fine scale MX-type precipitation observing at grain interiors. Elemental composition of precipitates described by EDS analysis

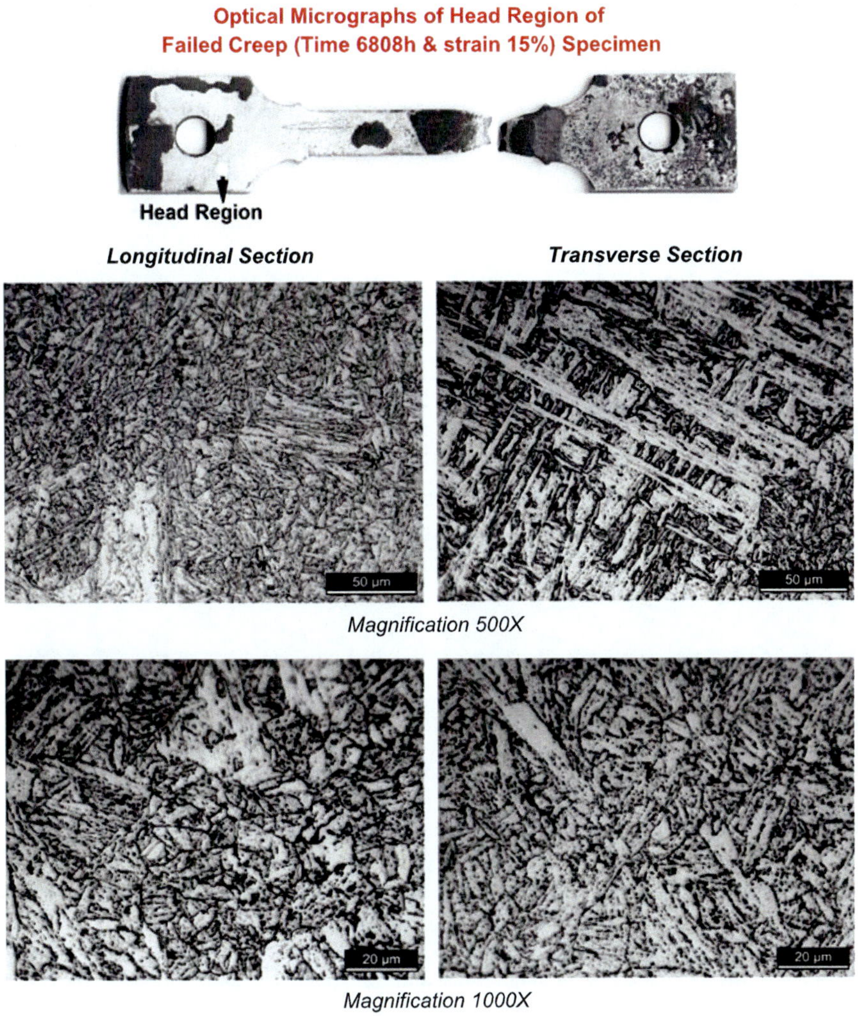

Fig. 4.35 Optical micrographs of **longitudinal and transverse sections** of P91 (9Cr-1Mo) steel after creep exposed at 550 °C/200 MPa for 6808 h, showing martensite matrix

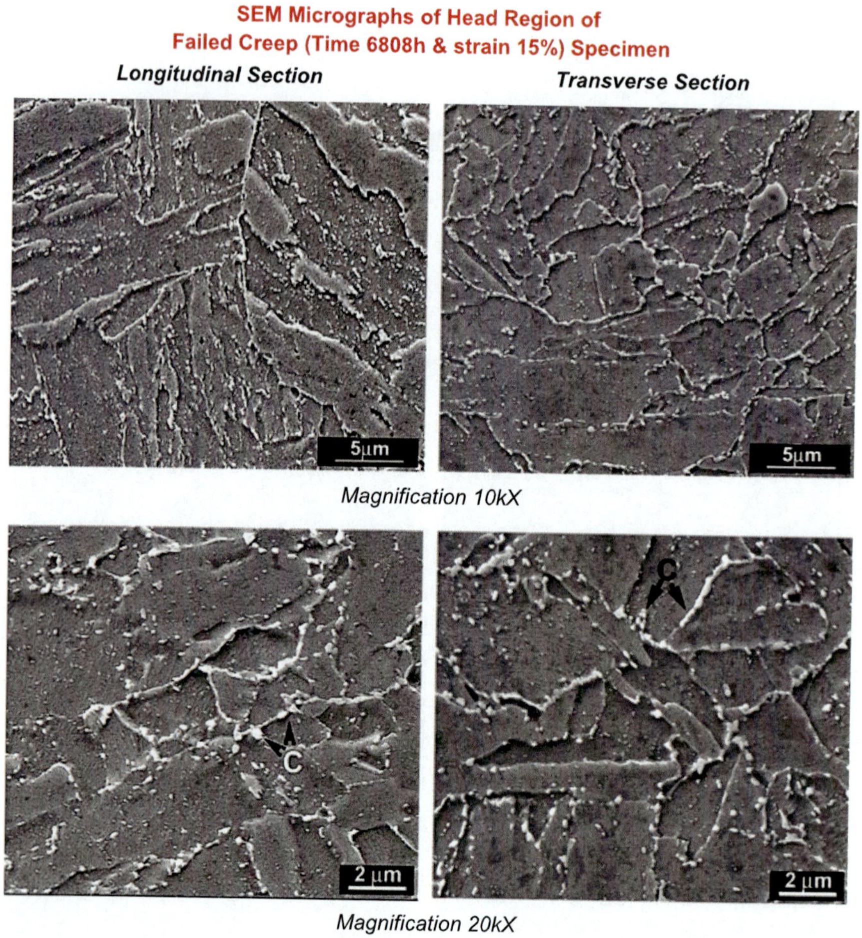

Fig. 4.36 SEM-SE micrographs of **longitudinal and transverse sections** of P91 (9Cr-1Mo) steel after creep exposed at 550 °C/200 MPa for 6808 h, showing carbide (C) particles distributed inhomogeneously in the martensitic matrix and grain boundary

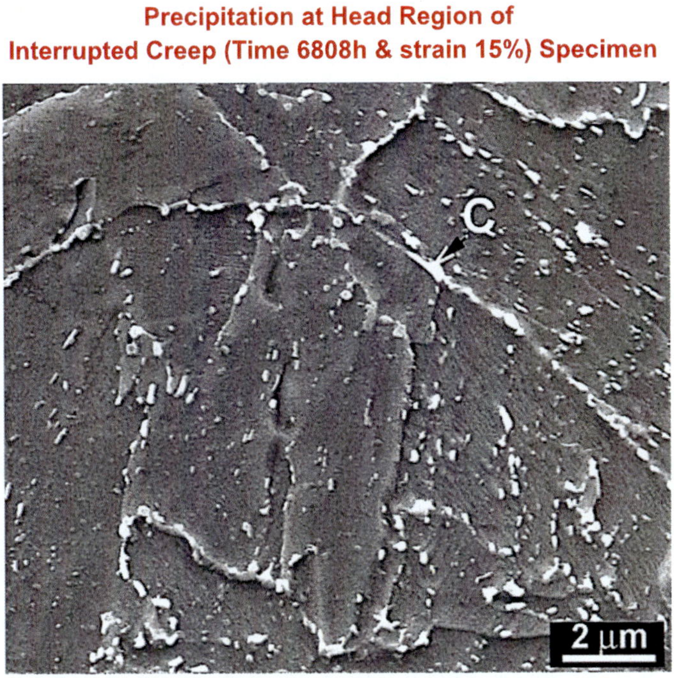

Fig. 4.37 SEM-SE micrograph (magnification 20kX) of ***longitudinal section*** of P91 (9Cr-1Mo) steel after creep exposed at 550 °C/200 MPa for 6808 h, showing $M_{23}C_6$ (point C) precipitates at grain interiors and grain boundaries of martensitic structures

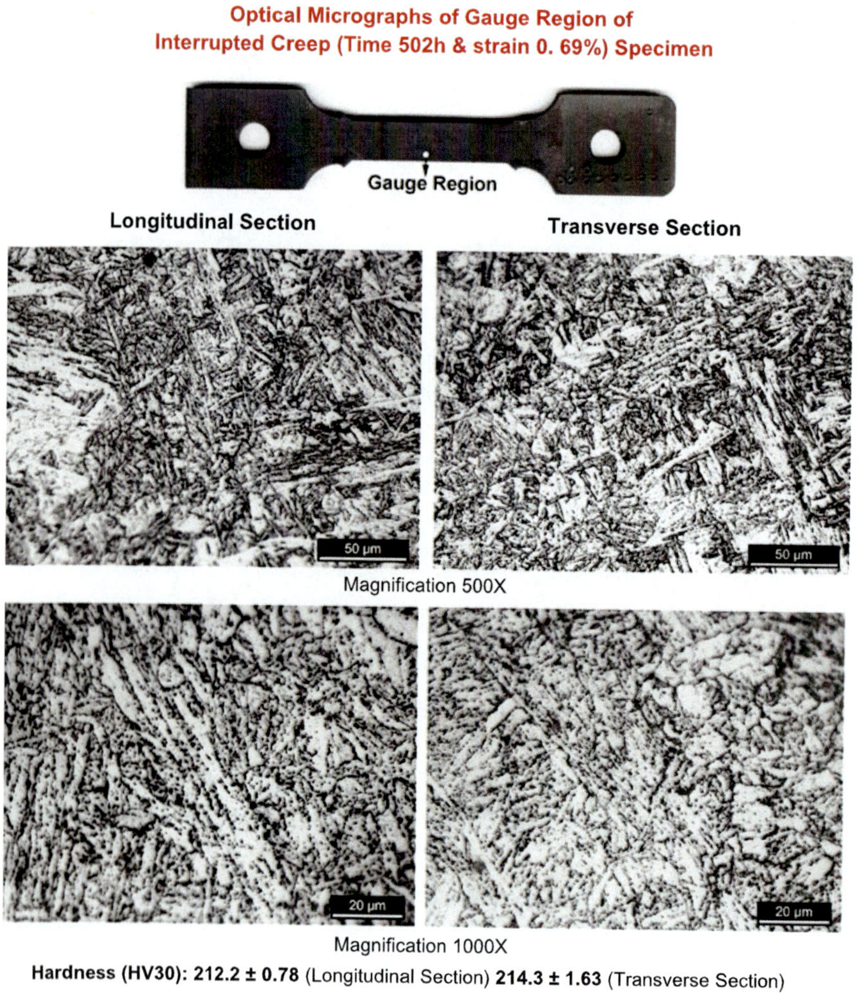

Fig. 4.38 Optical micrographs of ***longitudinal and transverse sections*** of P91 (9Cr-1Mo) steel after creep exposed at 600 °C/140 MPa for 502 h with a creep strain of 0.69%, showing martensitic structures

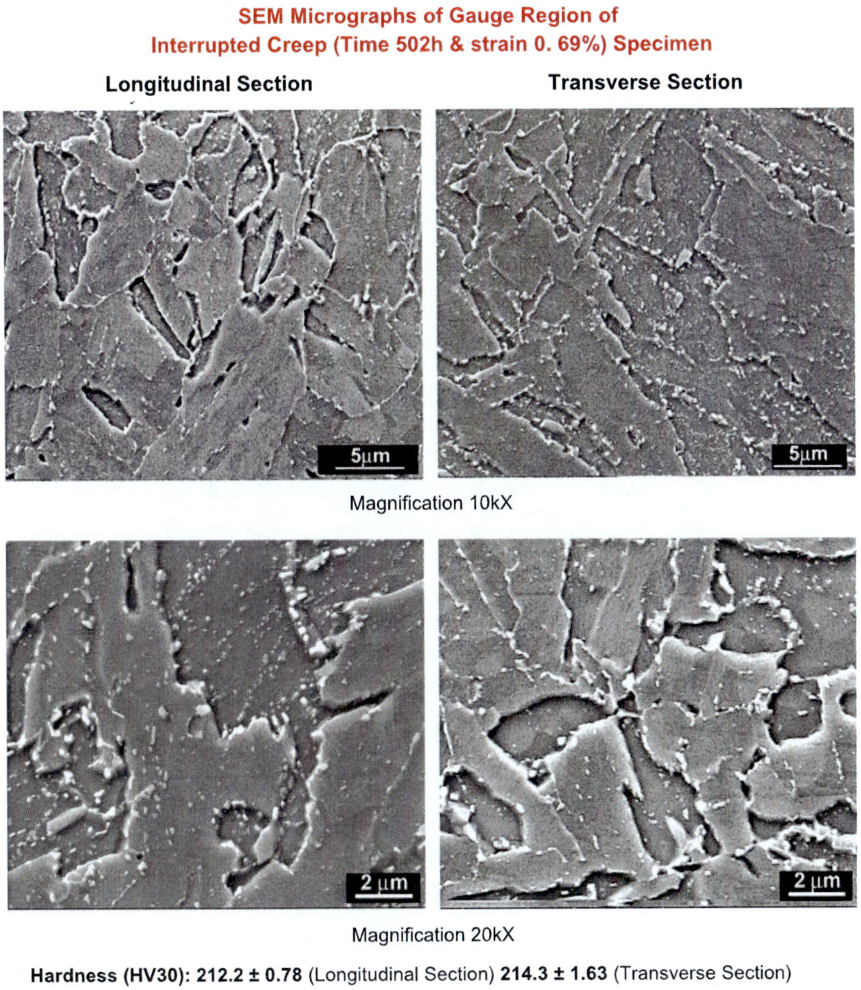

Fig. 4.39 SEM-SE micrographs of ***longitudinal and transverse sections*** of P91 (9Cr-1Mo) steel after creep exposed at 600 °C/140 MPa for 502 h with a creep strain of 0.69%, showing lath martensites

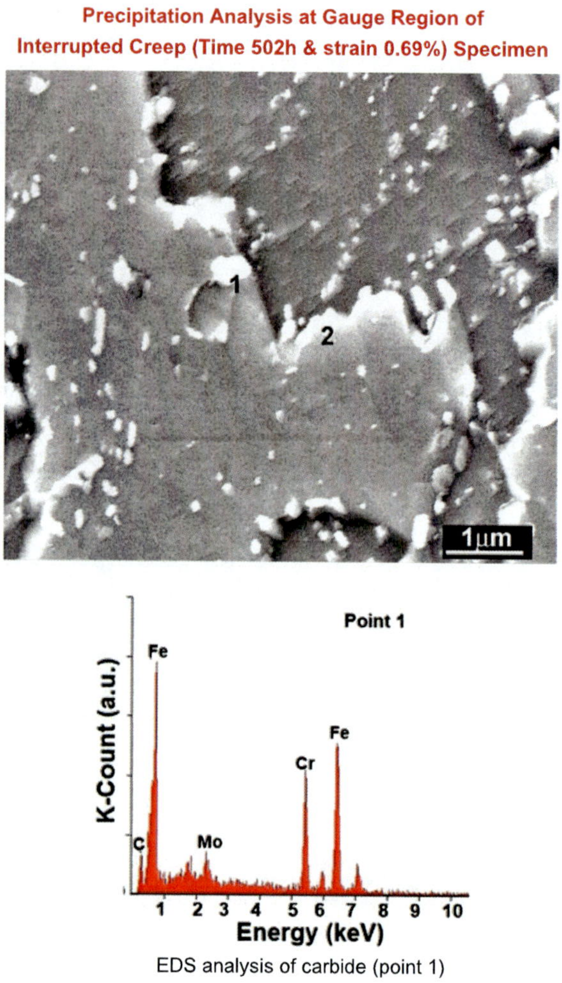

Fig. 4.40 SEM-SE micrograph (magnification 40kX) of ***longitudinal section*** of P91 (9Cr-1Mo) steel after creep exposed at 600 °C/140 MPa for 502 h with a creep strain of 0.69%, showing coarse Cr-rich $M_{23}C_6$ carbides (points 1 and 2) at grain boundaries

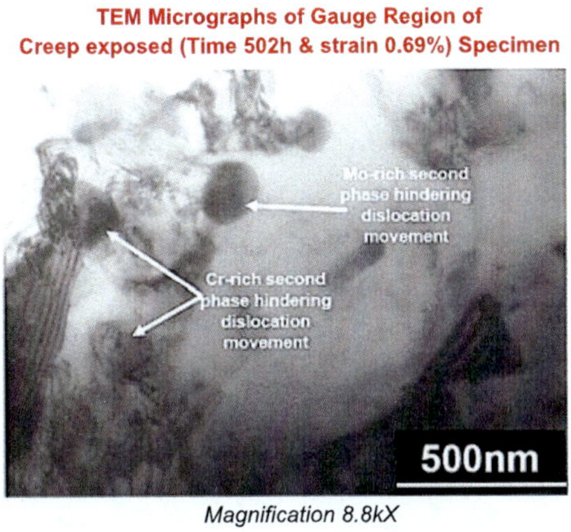

Fig. 4.41 TEM-BF image of <u>gauge region</u> at ***transverse section*** of P91(9Cr-1Mo) steel after creep exposed at 600 °C/140 MPa for 502 h with a creep strain of 0.69%, explaining second-phase precipitation which occurred near grain boundary triple points or at subgrain boundaries. Physical appearance of the same, i.e., the rounding of, edge of rectangular precipitates, or near elliptical shape indicated substantial growth under stress and temperature to minimize the surface free energy. The geometry endorsed incoherency of the second phase with respect to the matrix. Owing to the incoherency nature of second phase, the dislocations could not bypass nor cut through them, rather they have been strongly pinned by the precipitate. Thus around the precipitate, qualitatively the dislocation density became more with respect to rest of the surrounding area

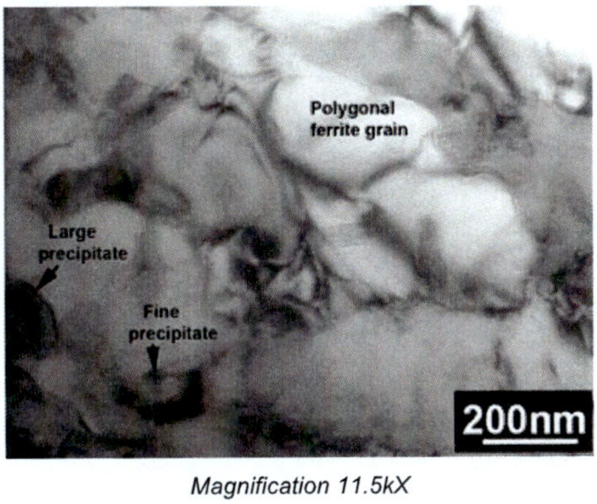

Fig. 4.42 TEM-BF image of <u>gauge region</u> at ***transverse section*** of P91(9Cr-1Mo) steel after creep exposed at 600 °C/140 MPa for 502 h with a creep strain of 0.69%, explaining clustered dislocations moved and around the barriers like grain boundary and precipitate under stress at elevated temperature. Grain body appeared clean, which referred to rearrangement of dislocations once again. The precipitate dimension was also heterogeneous in nature, which indicated that preferred orientation of precipitates with respect to matrix played a key role in the subsequent growth of the same during deformation

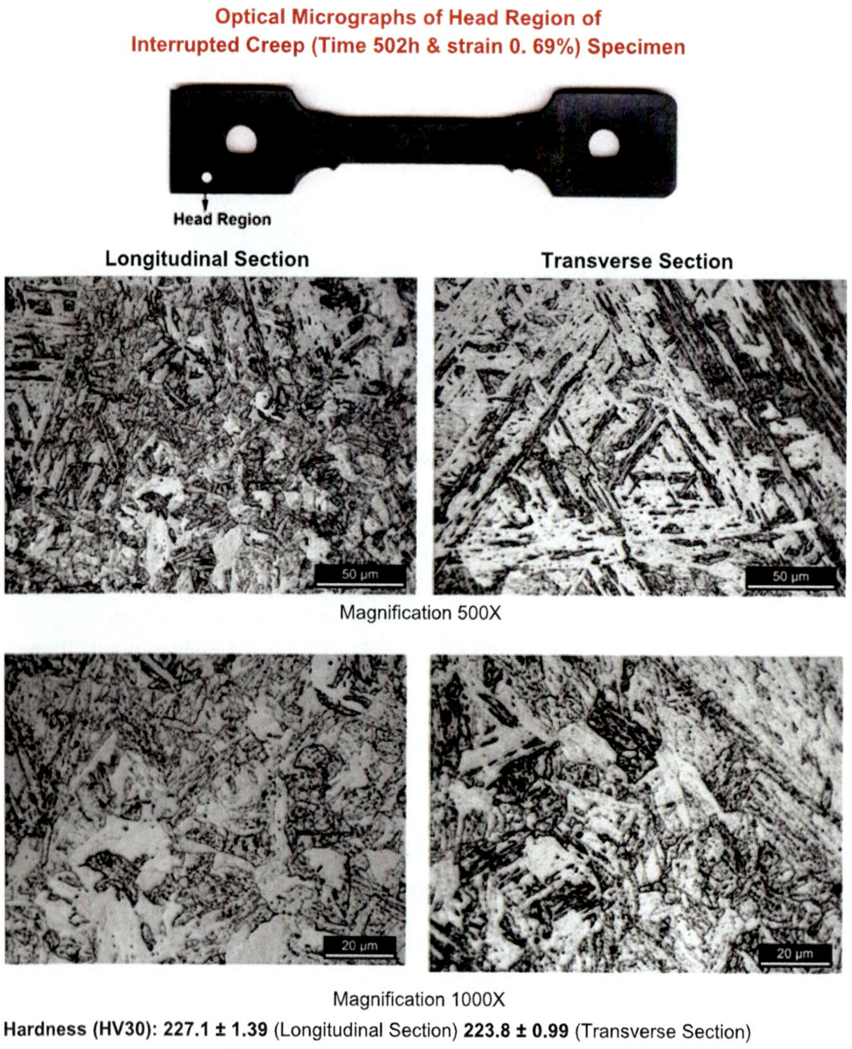

Fig. 4.43 Optical micrographs of ***longitudinal and transverse sections*** of P91 (9Cr-1Mo) steel after creep exposed at 600 °C/140 MPa for 502 h with a creep strain of 0.69%, showing martensitic structures. Some discrete martensite phase distributed like islands

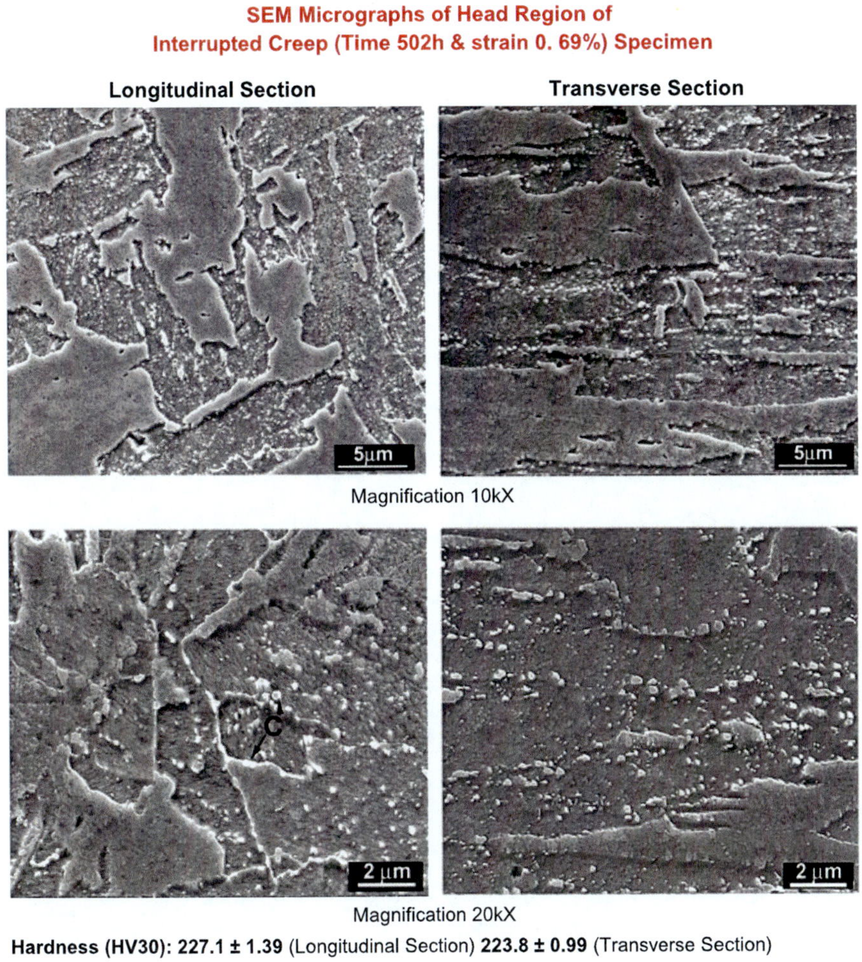

Fig. 4.44 SEM-SE micrographs of ***longitudinal and transverse sections*** of P91 (9Cr-1Mo) steel after creep exposed at 600 °C/140 MPa for 502 h with a creep strain of 0.69%, showing martensitic structures. Some discrete blocky martensites distributed like islands and carbides are in grain interiors and grain boundaries

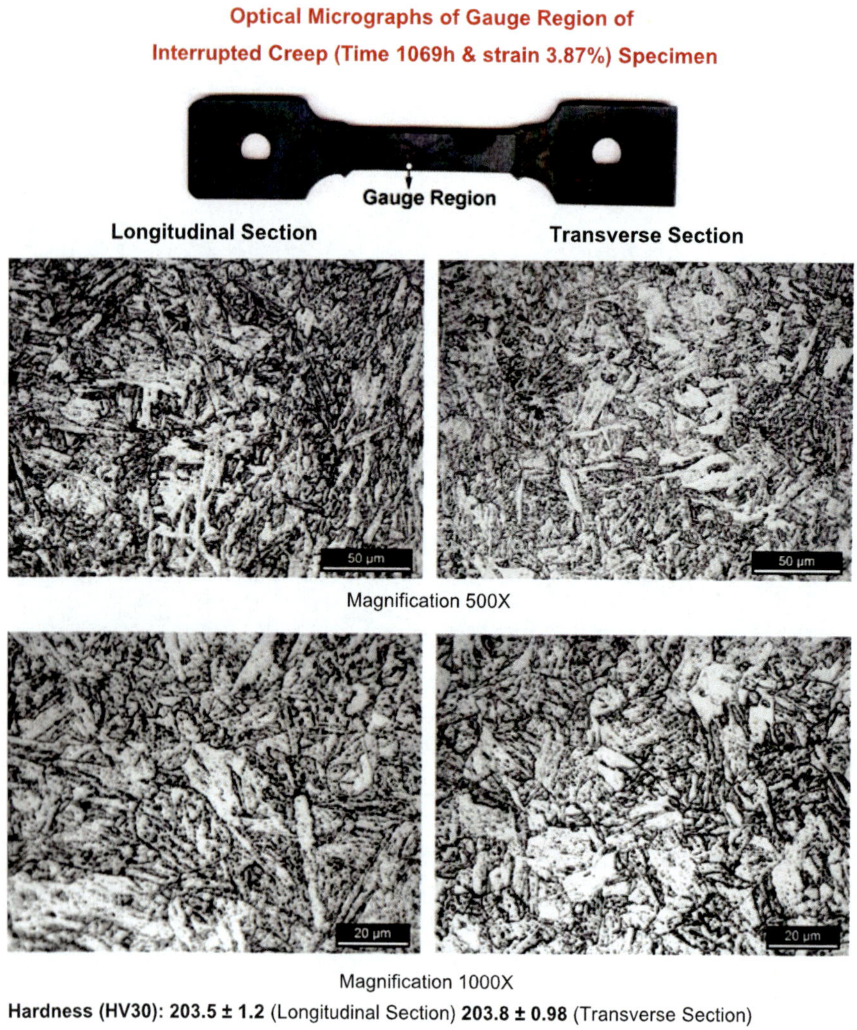

Fig. 4.45 Optical micrographs of ***longitudinal and transverse sections*** of P91 (9Cr-1Mo) steel after creep exposed at 600 °C/140 MPa for 1069 h with a creep strain of 3.87%, showing martensitic structures

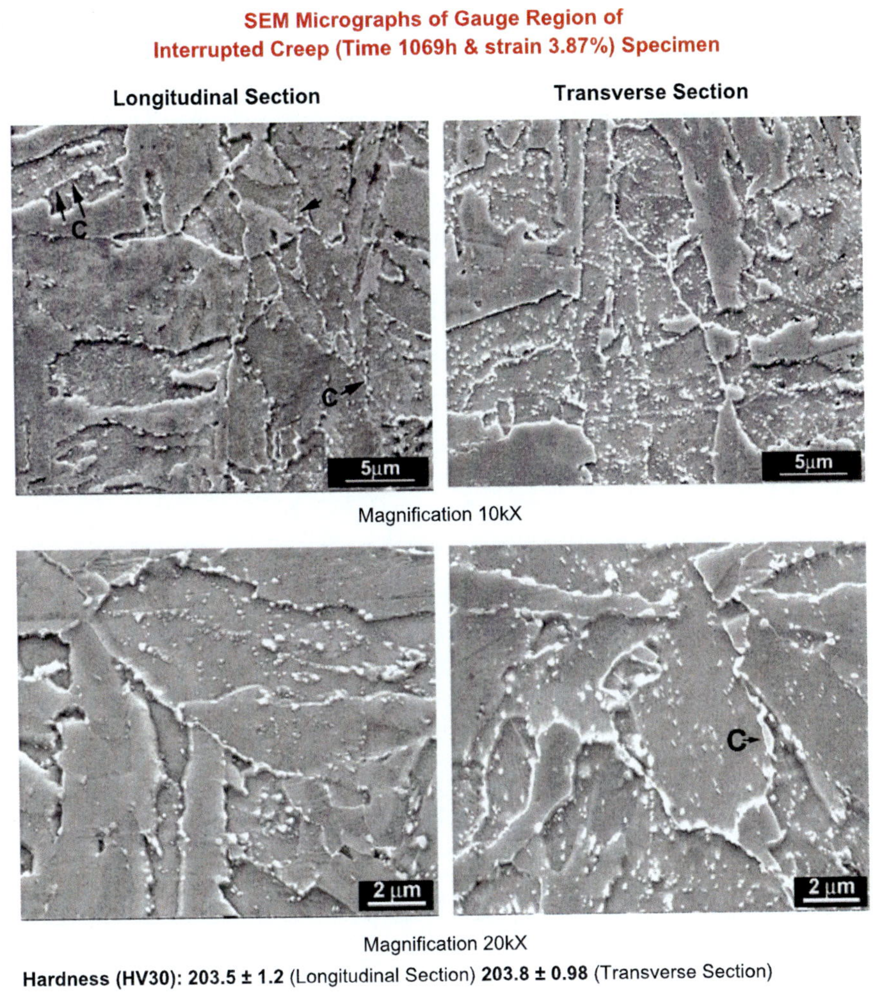

Fig. 4.46 SEM-SE micrographs of ***longitudinal and transverse sections*** of P91 (9Cr-1Mo) steel after creep exposed at 600 °C/140 MPa for 1069 h with a creep strain of 3.87%, showing carbide (C) precipitates distributed as a chain (double black arrows) and discrete way at grain boundaries and grain interiors

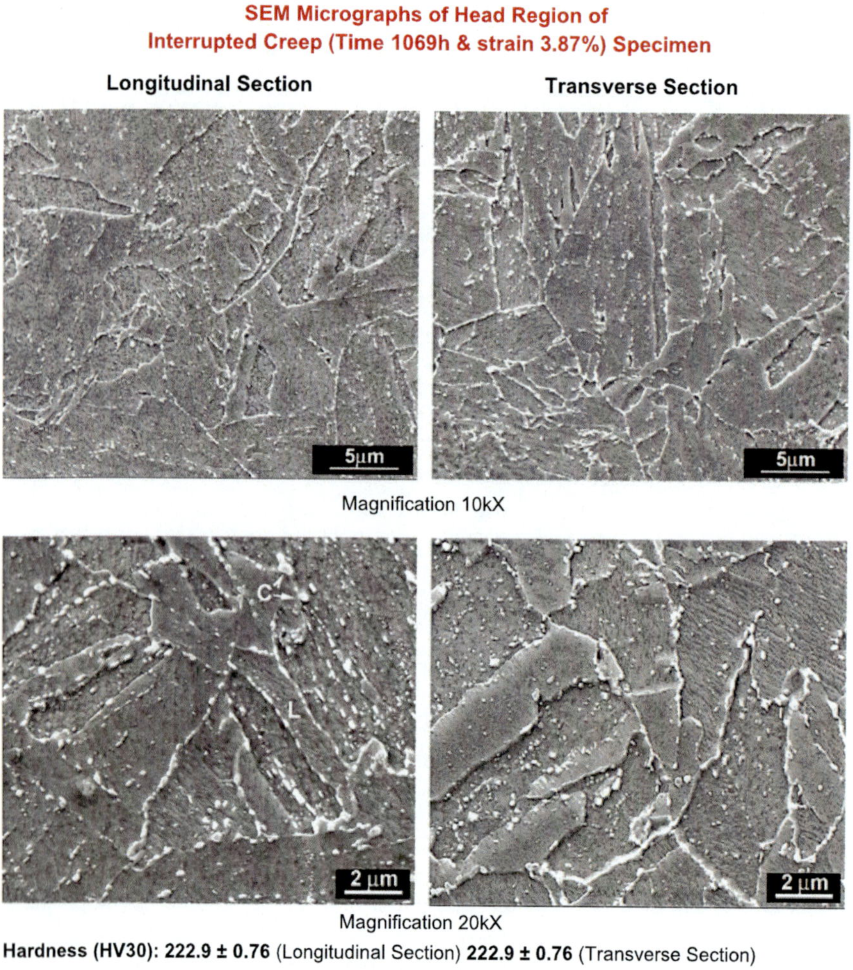

Fig. 4.47 SEM-SE micrographs of ***longitudinal and transverse sections*** of P91 (9Cr-1Mo) steel after creep exposed at 600 °C/140 MPa for 1069 h with a creep strain of 3.87%, showing chain-like distributed carbides at lath (L) martensitic boundaries

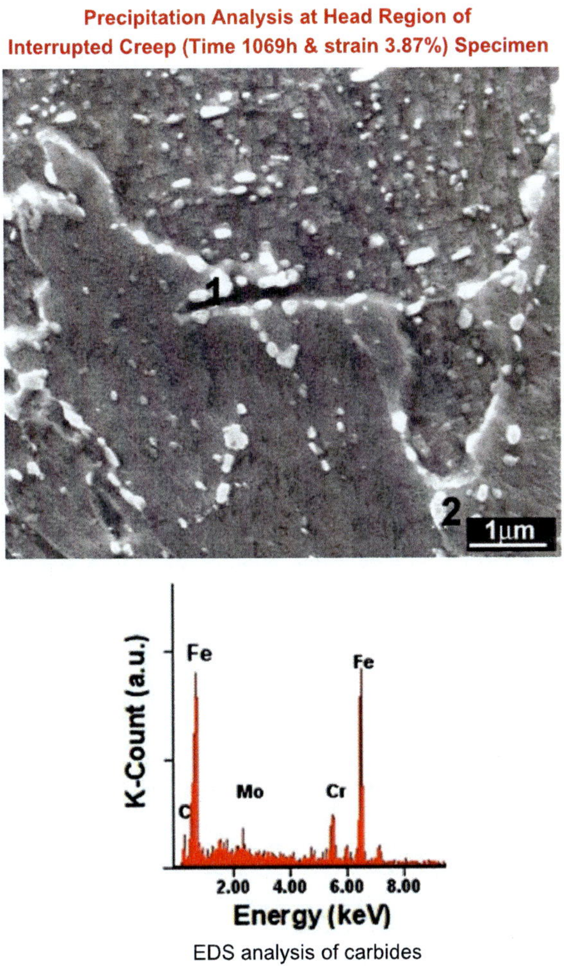

Fig. 4.48 SEM-SE micrograph (magnification 40kX) of ***longitudinal sections*** of P91 (9Cr-1Mo) steel after creep exposed at 600 °C/140 MPa for 1069 h with a creep strain of 3.87%, showing coarse Cr-rich carbides at grain boundaries

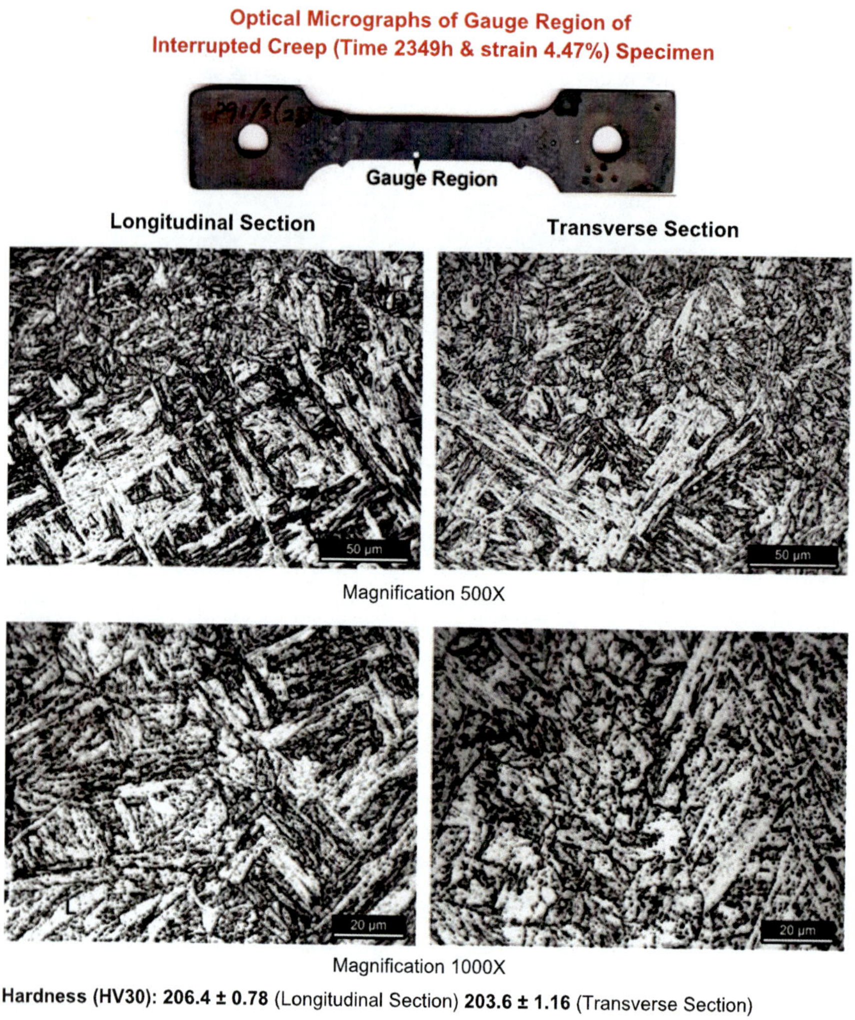

Fig. 4.49 Optical micrographs of ***longitudinal and transverse sections*** of P91 (9Cr-1Mo) steel after creep exposed at 600 °C/140 MPa for 2349 h with a creep strain of 4.47%, showing martensitic structures

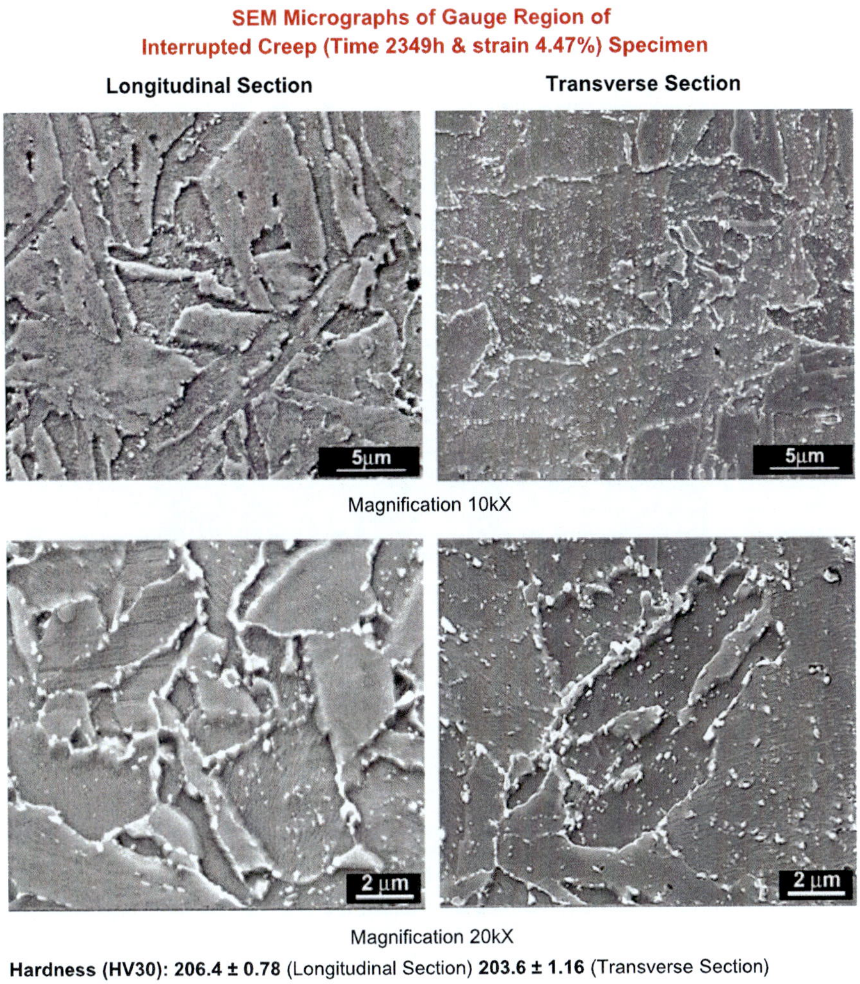

Fig. 4.50 SEM-SE micrographs of ***longitudinal and transverse sections*** of P91 (9Cr-1Mo) steel after creep exposed at 600 °C/140 MPa for 2349 h with a creep strain of 4.47%, showing distributions of carbides at grain interiors and boundaries of martensitic laths

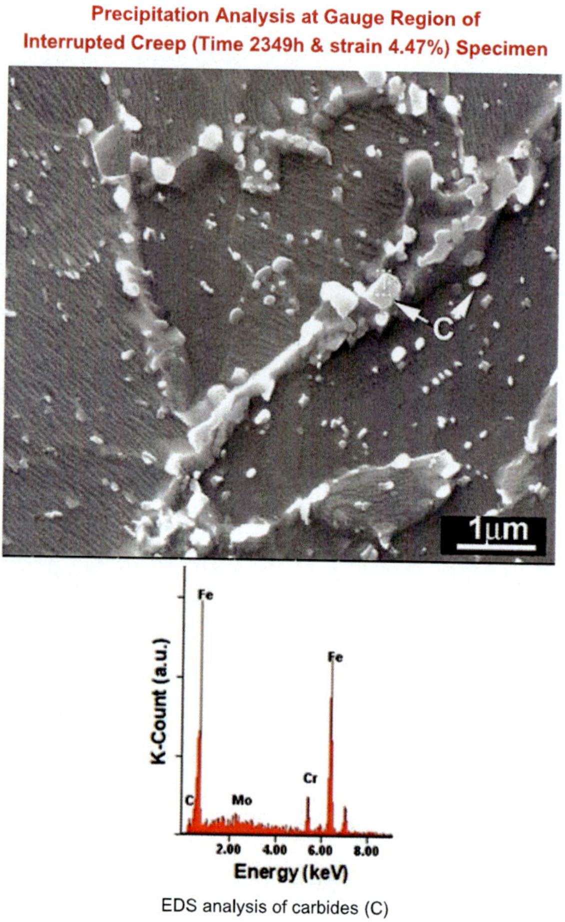

Fig. 4.51 SEM-SE micrograph (magnification 40kX) of ***transverse sections*** of P91 (9Cr-1Mo) steel after creep exposed at 600 °C/140 MPa for 2349 h with a creep strain of 4.47%, showing coarse (Fe, Cr, Mo)-rich carbides (C) at grain boundaries

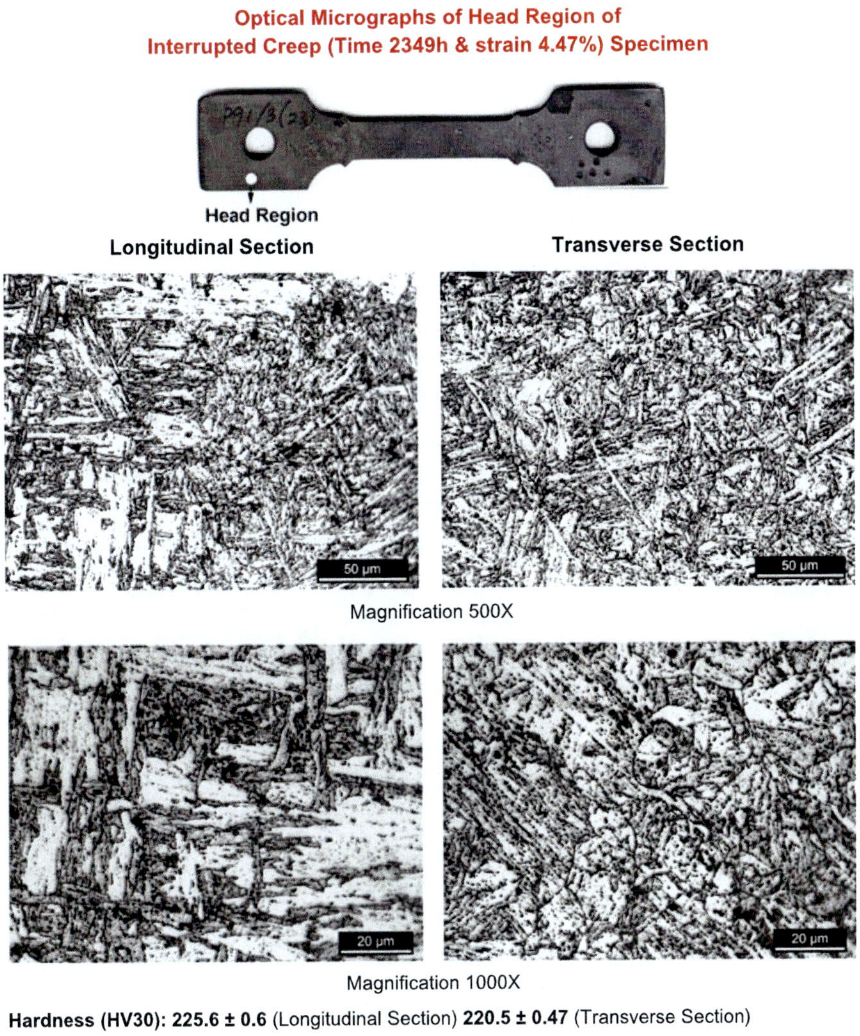

Fig. 4.52 Optical micrographs of ***longitudinal and transverse sections*** of P91 (9Cr-1Mo) steel after creep exposed at 600 °C/140 MPa for 2349 h with a creep strain of 4.47%, showing martensitic structures

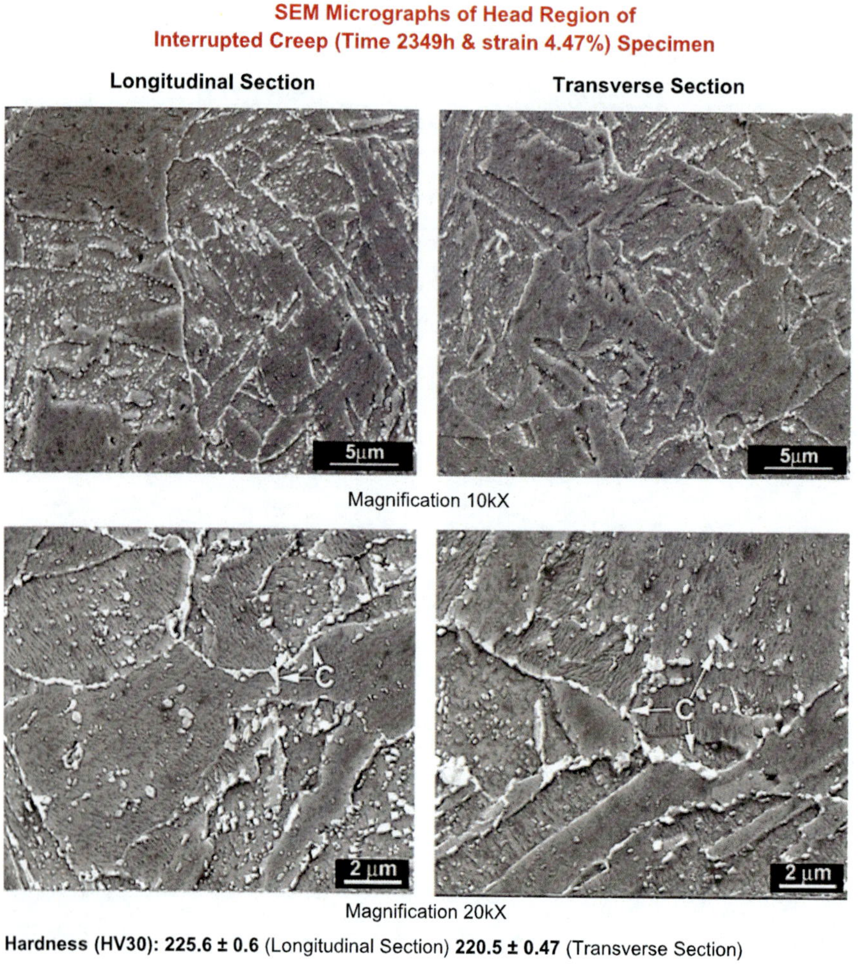

Fig. 4.53 SEM-SE micrographs of ***longitudinal and transverse sections*** of P91 (9Cr-1Mo) steel after creep exposed at 600 °C/140 MPa for 2349 h with a creep strain of 4.47%, showing the distribution of carbide (C) precipitates at the grain boundaries and grain interiors of martensitic structures

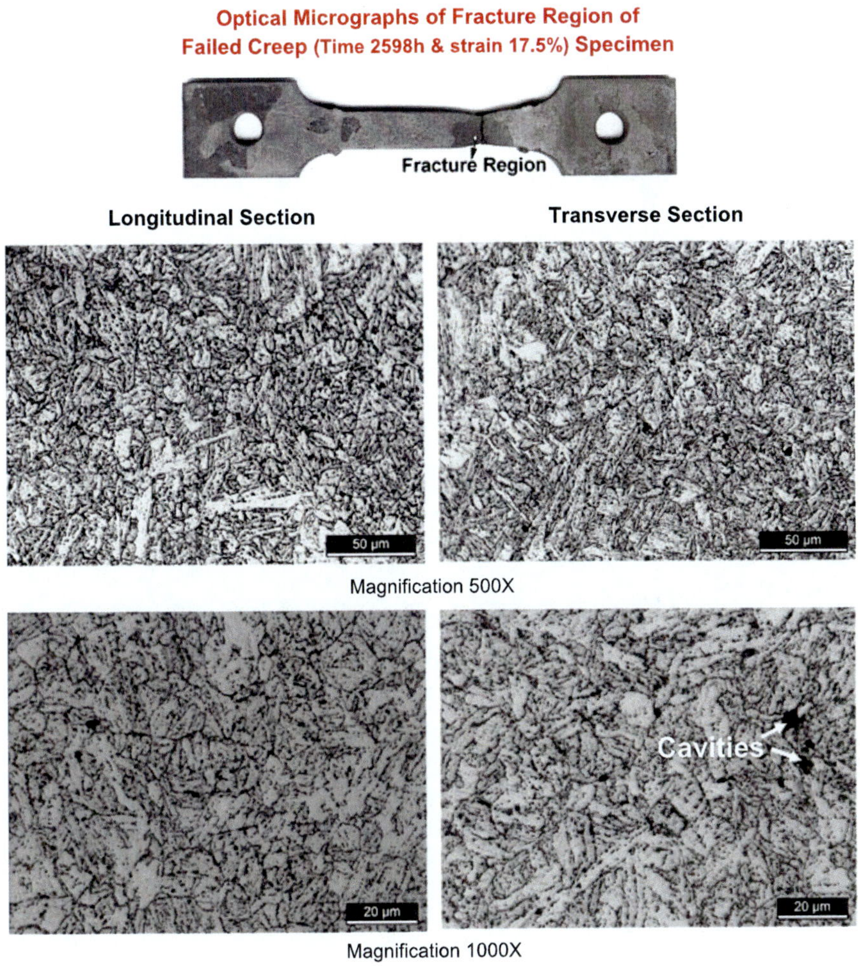

Fig. 4.54 Optical micrographs of ***longitudinal and transverse sections*** of P91 (9Cr-1Mo) steel after creep failed at 600 °C/140 MPa for 2598 h with a creep strain of 17.5%, showing martensitic structures along with inhomogeneous distribution of cavities

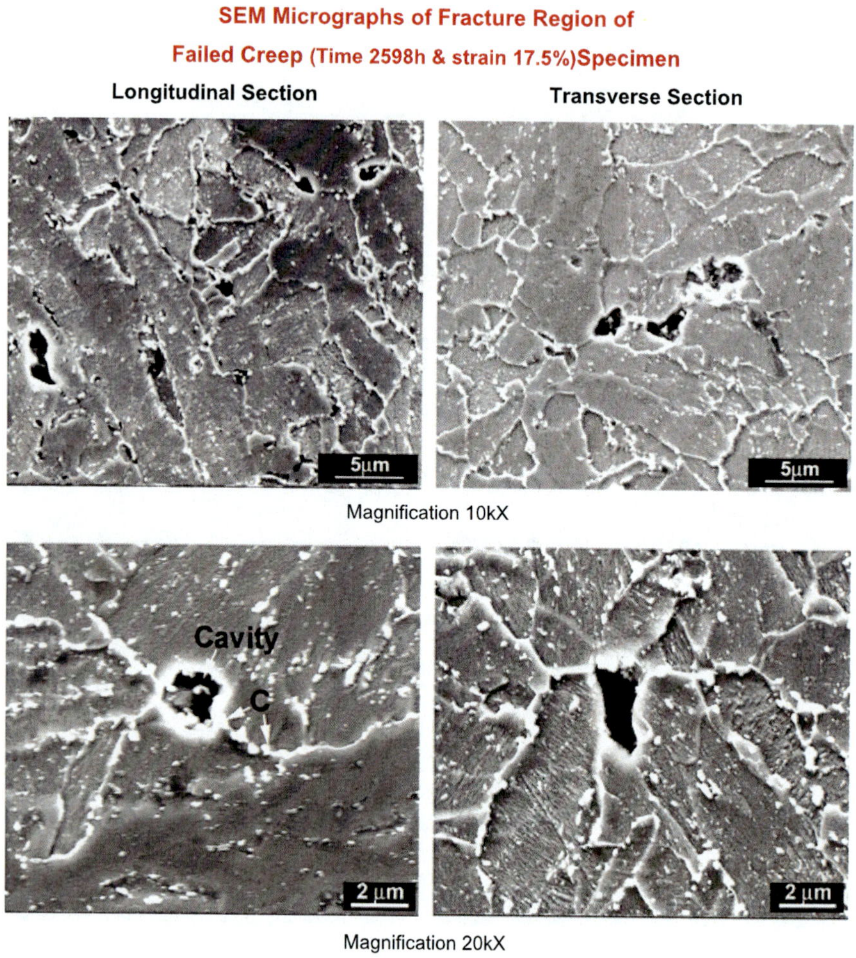

Fig. 4.55 SEM-SE micrographs of ***longitudinal and transverse sections*** of P91 (9Cr-1Mo) steel after creep failed at 600 °C/140 MPa for 2598 h with a creep strain of 17.5%, showing distribution of carbides (C) and cavities at grain boundaries of martensitic structures

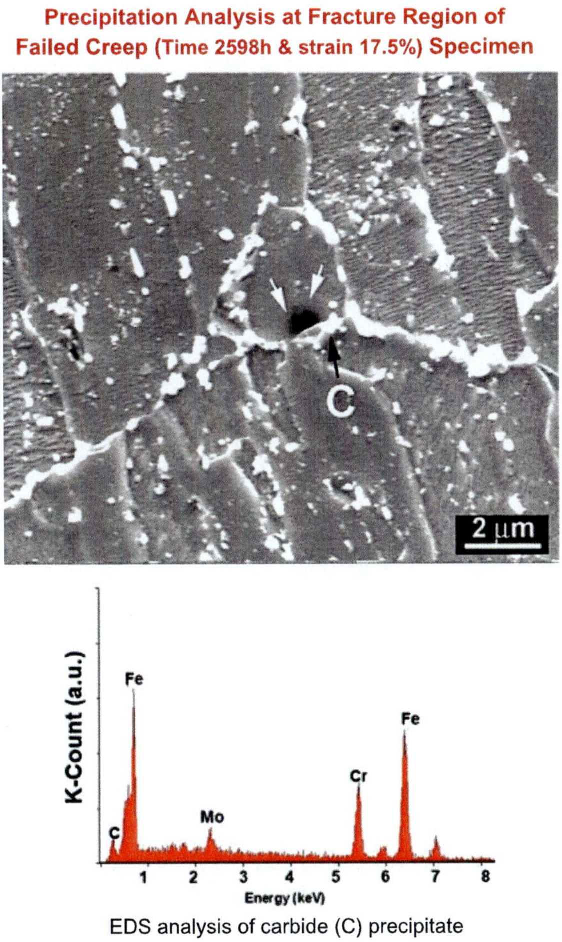

Fig. 4.56 SEM-SE micrograph (magnification 20kX) of ***longitudinal section*** of P91 (9Cr-1Mo) steel after creep exposed at 600 °C/140 MPa for 2598 h with a creep strain of 17.5%, showing Cr-rich carbides (C) at grain boundaries and the initiation of cavities (by white arrows) at that place

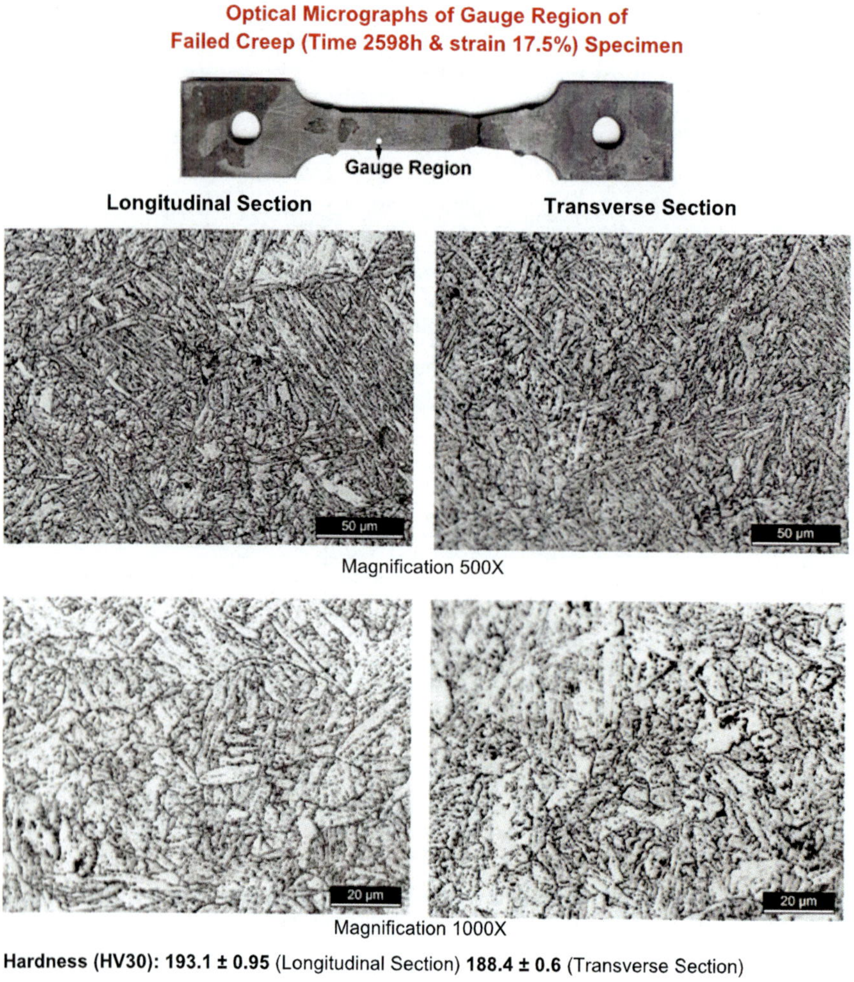

Fig. 4.57 Optical micrographs of ***longitudinal and transverse sections*** of P91 (9Cr-1Mo) steel after creep failed at 600 °C/140 MPa for 2598 h with a creep strain of 17.5%, showing martensitic structures

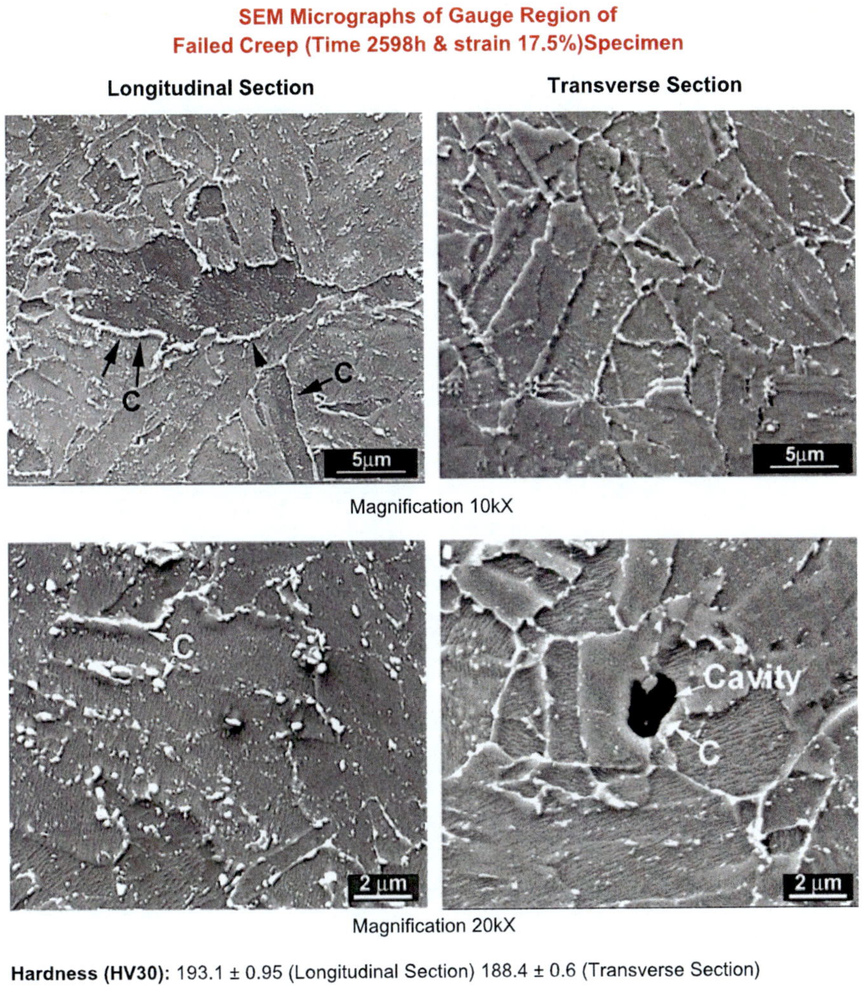

Fig. 4.58 SEM-SE micrographs of ***longitudinal and transverse sections*** of P91 (9Cr-1Mo) steel after creep failed at 600 °C/140 MPa for 2598 h with a creep strain of 17.5%, the distribution of carbides (C) and cavities at grain boundaries and interiors of martensitic structures

TEM Micrographs of Gauge Region of Failed Creep (Time 2598h & strain 17.5%) Specimen

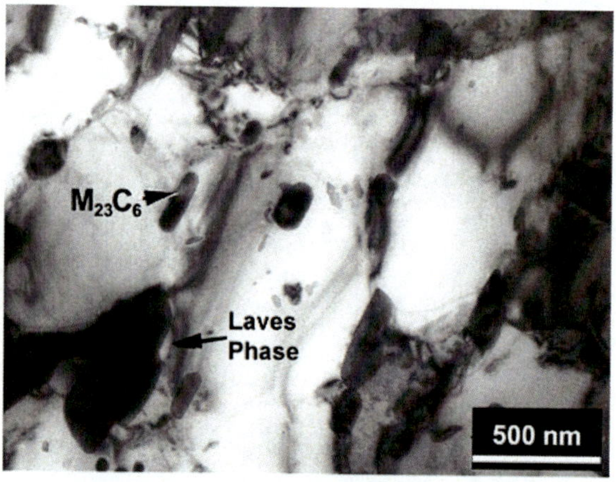

Magnification 11.5kX

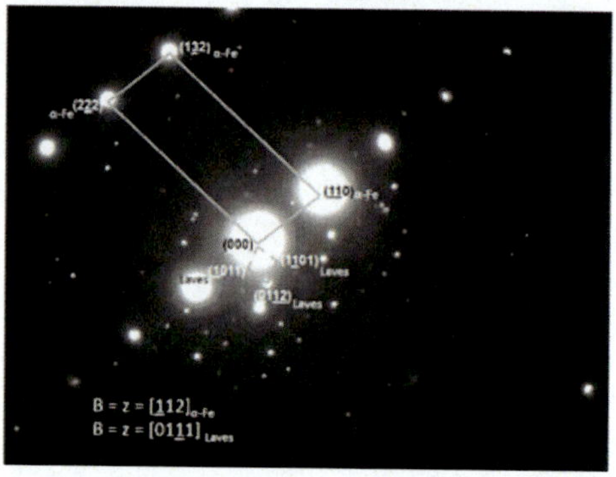

SAED pattern of Laves phase

Fig. 4.59 TEM-BF image of <u>*gauge region*</u> at ***transverse section*** of P91(9Cr-1Mo) steel after creep failed at 600 °C/140 MPa for 2598 h with a creep strain of 17.5%, showing coarse irregular-shaped Laves phase (with SAED pattern) pinning at grain boundary and parallelepiped-shaped $M_{23}C_6$ carbides at grain interiors

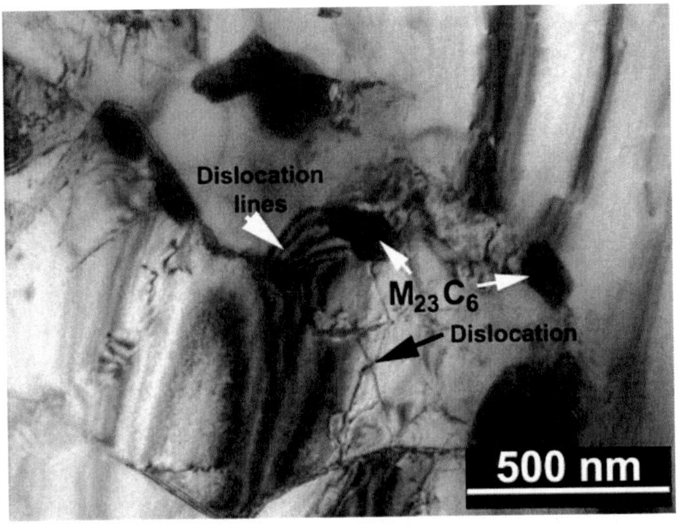

Magnification 15kX

Fig. 4.60 TEM-BF image of *gauge region* at ***transverse section*** of P91(9Cr-1Mo) steel after creep failed at 600 °C/140 MPa for 2598 h with a creep strain of 17.5%, showing $M_{23}C_6$ carbides interacting to dislocations and precipitate coagulation to exhibit increment in size

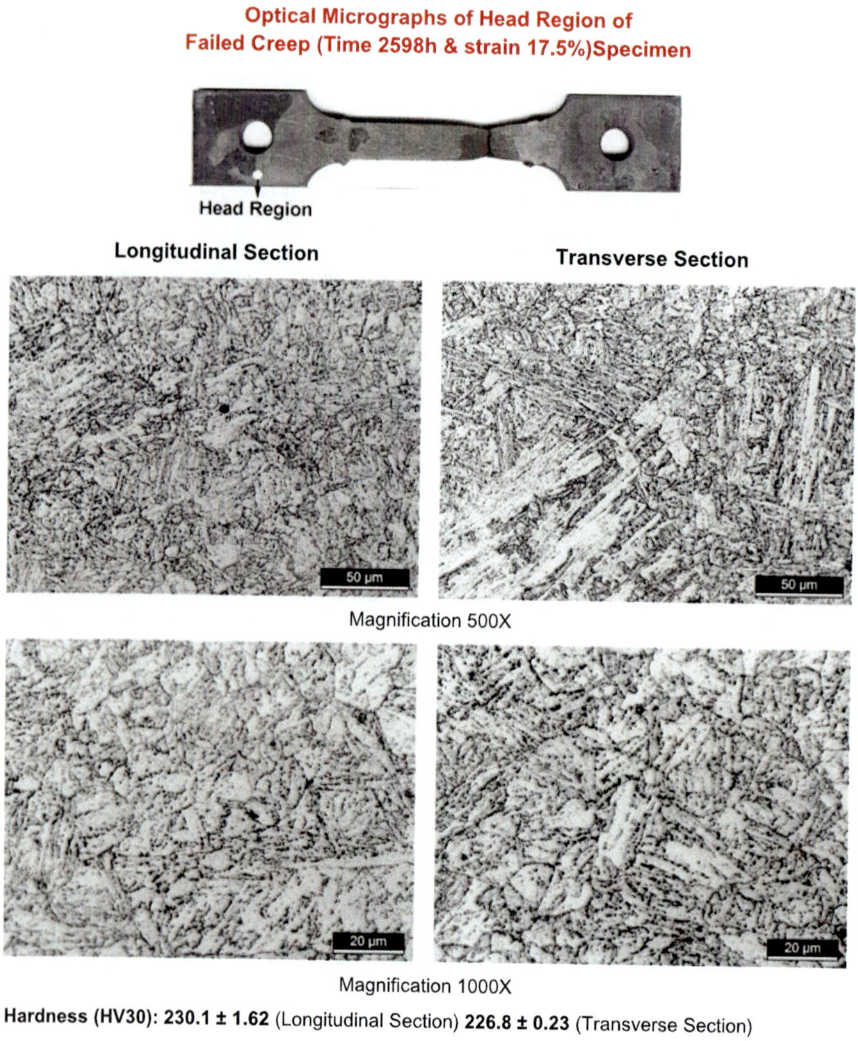

Fig. 4.61 Optical micrographs of ***longitudinal and transverse sections*** of P91 (9Cr-1Mo) steel after creep failed at 600 °C/140 MPa for 2598 h with a creep strain of 17.5%, showing martensitic structures

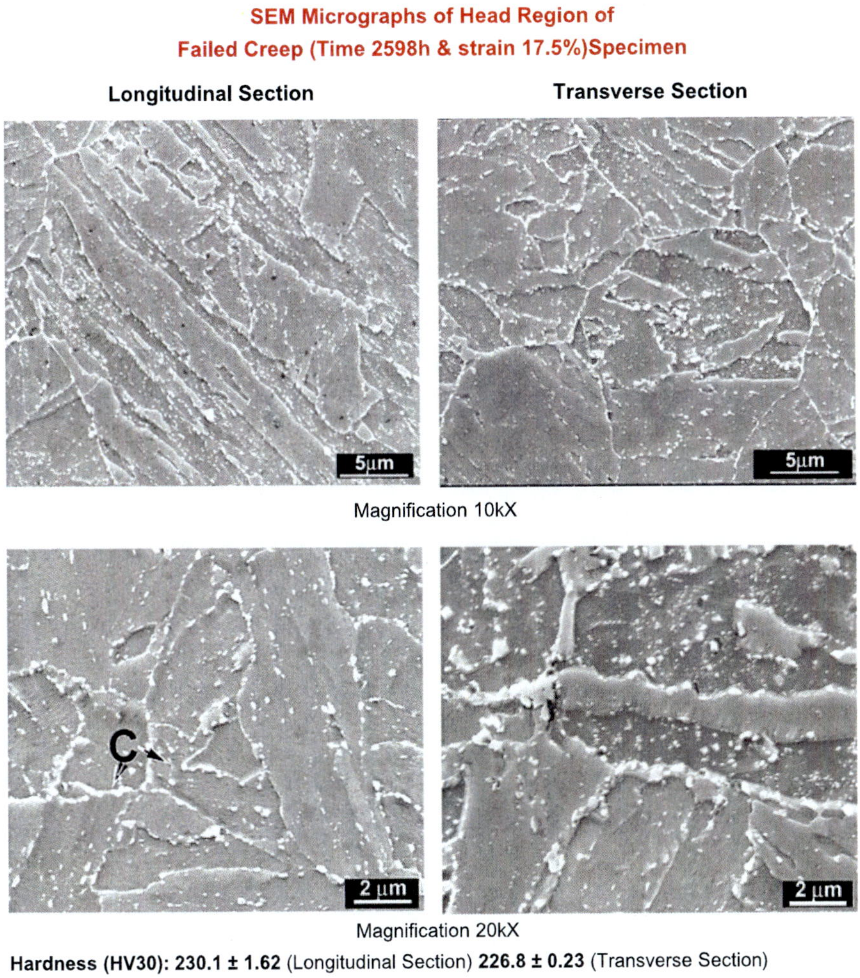

Fig. 4.62 SEM-SE micrographs of ***longitudinal and transverse sections*** of P91 (9Cr-1Mo) steel after creep failed at 600 °C/140 MPa for 2598 h with a creep strain of 17.5%, showing the distribution of coarse carbides (C) at grain boundaries and fine carbides at grain interiors of martensitic structures

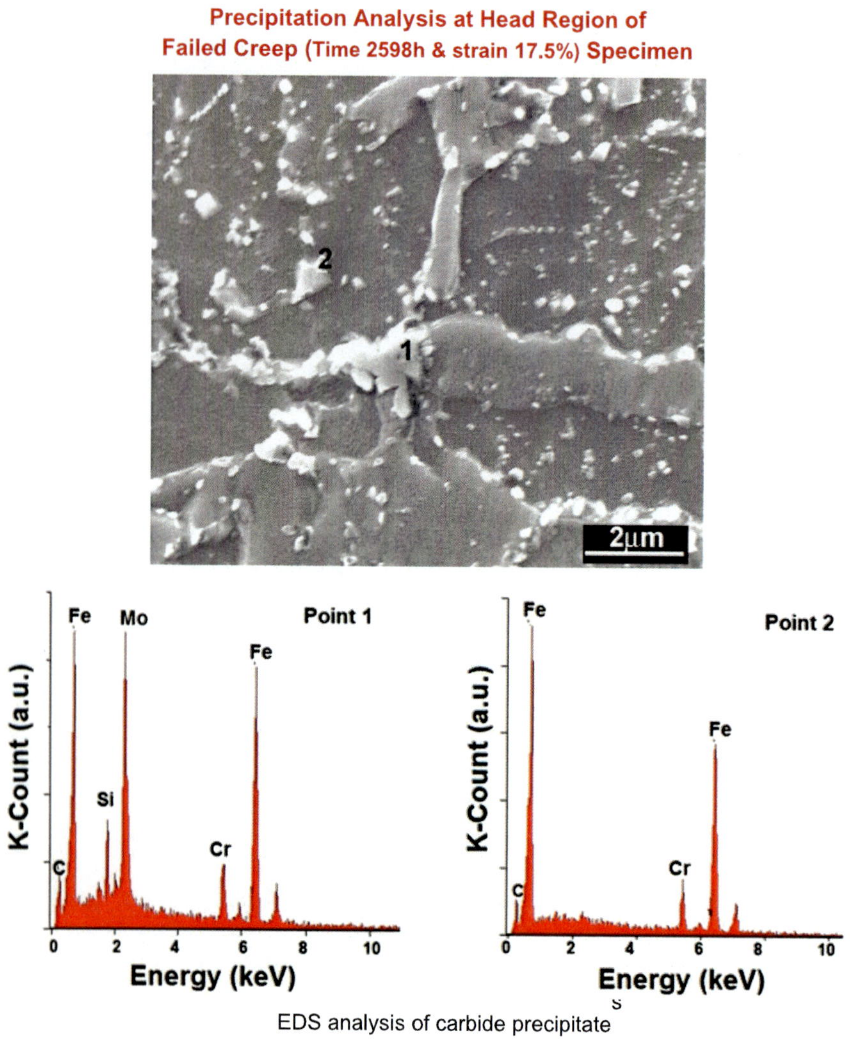

Fig. 4.63 SEM-SE micrograph (magnification 30kX) of ***transverse section*** of P91 (9Cr-1Mo) steel after creep exposed at 600 °C/140 MPa for 2598 h with a creep strain of 17.5%, showing Laves phase (point 1) and Cr-rich carbides (point 2) at grain boundaries and grain interiors

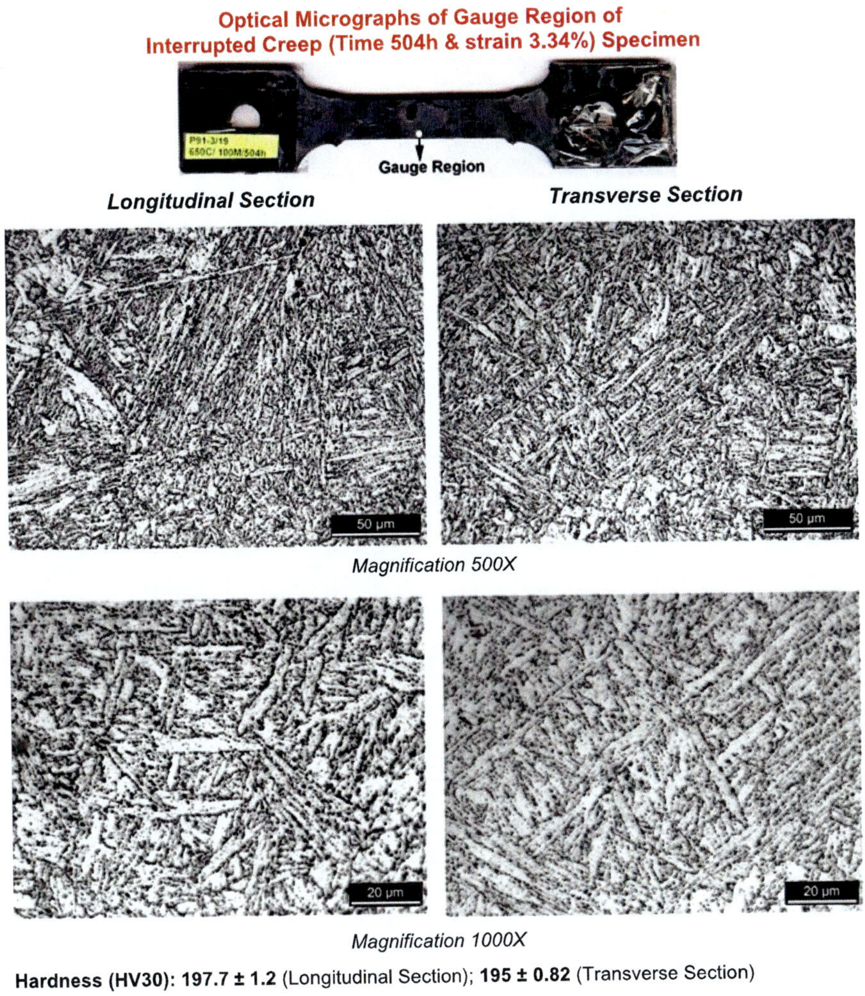

Fig. 4.64 Optical micrographs of **longitudinal and transverse sections** of P91 (9Cr-1Mo) steel after creep exposed at 650 °C/100 MPa for 504 h, showing lath martensitic structures

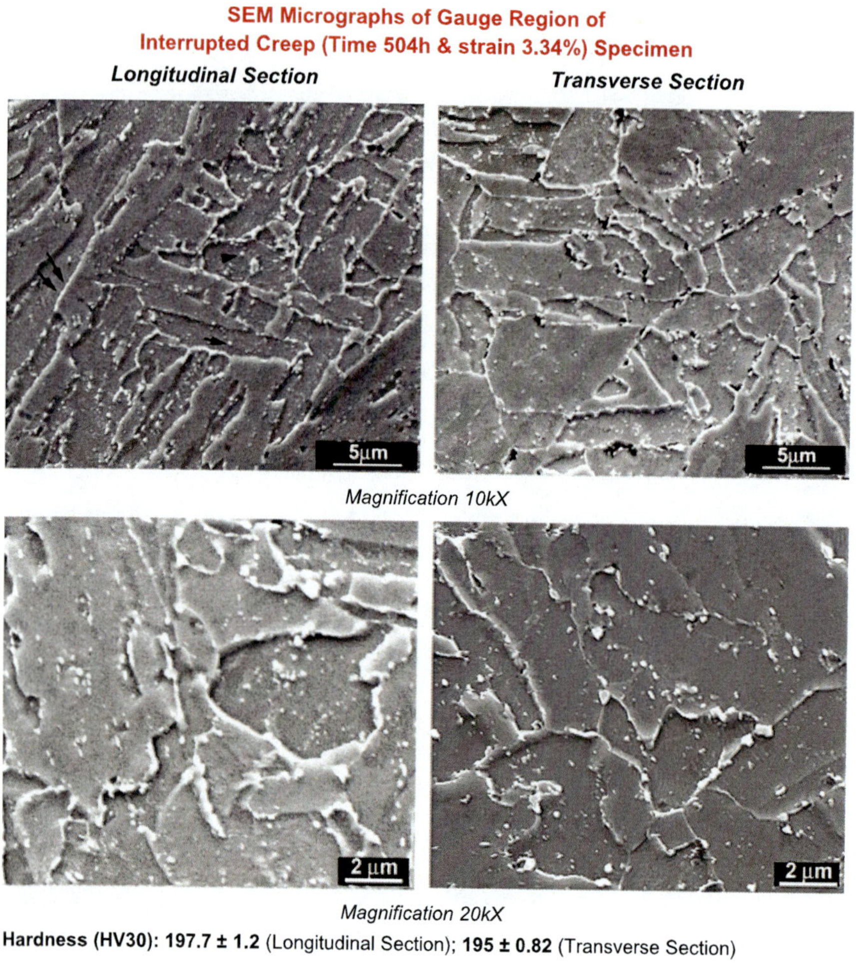

Fig. 4.65 SEM-SE micrographs of **longitudinal and transverse sections** of P91 (9Cr-1Mo) steel after creep exposed at 650 °C/100 MPa for 504 h, explaining the distribution of carbides (black arrows) at grain interiors and boundaries of martensitic structures

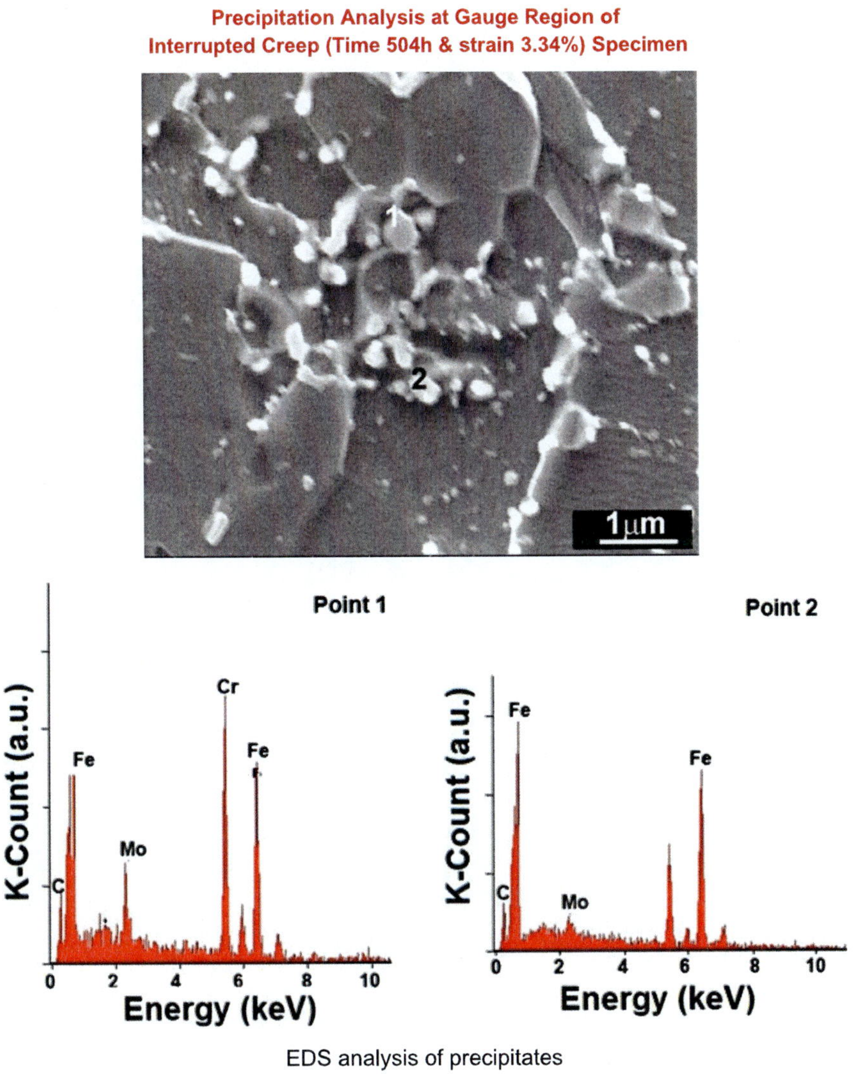

Fig. 4.66 SEM-SE micrograph (magnification 20kX) of ***transverse sections*** of P91 (9Cr-1Mo) steel after creep exposed at 650 °C/100 MPa for 504 h, showing Cr-rich (Fe, Cr, Mo) carbide (points 1 and 2) precipitates at grain boundaries of martensitic structures

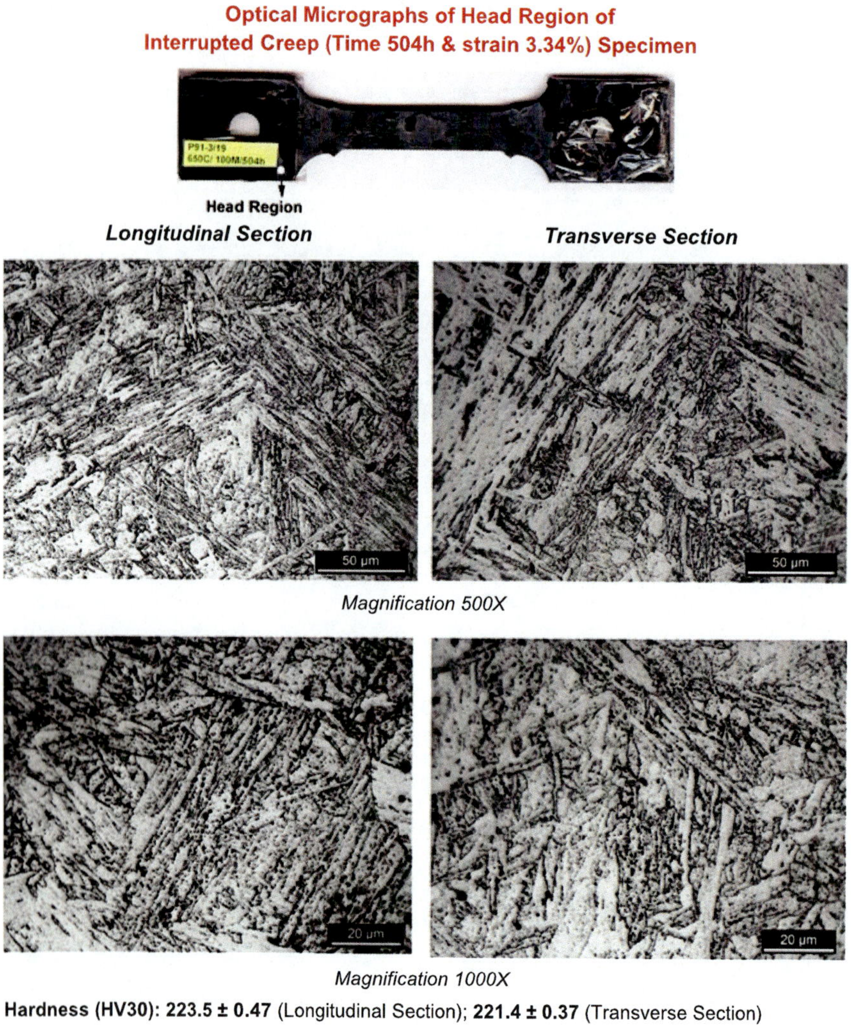

Fig. 4.67 Optical micrographs of **longitudinal and transverse sections** of P91 (9Cr-1Mo) steel after creep exposed at 650 °C/100 MPa for 504 h, showing martensitic structures

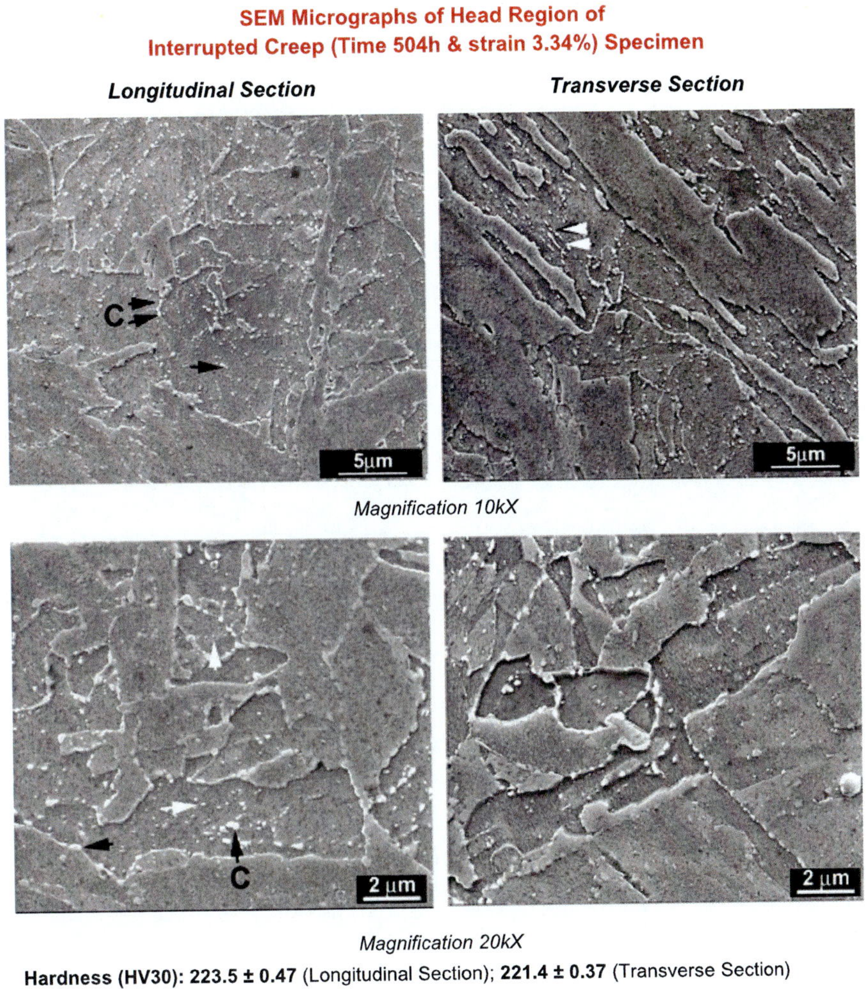

Fig. 4.68 SEM-SE micrographs of **longitudinal and transverse sections** of P91 (9Cr-1Mo) steel after creep exposed at 650 °C/100 MPa for 504 h, showing the distribution of coarse (black arrows) and fine (white arrows) carbides (C) at martensitic matrix

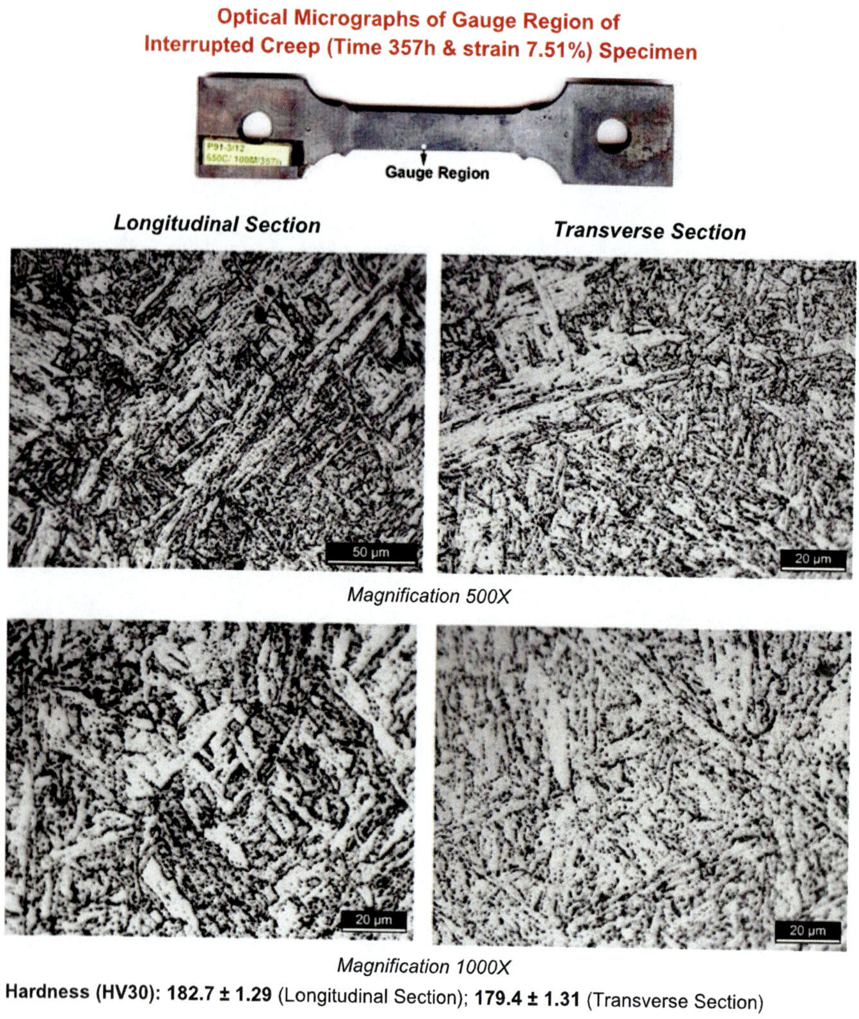

Fig. 4.69 Optical micrographs of **longitudinal and transverse sections** of P91 (9Cr-1Mo) steel after creep exposed at 650 °C/100 MPa for 357 h, explaining martensitic structures

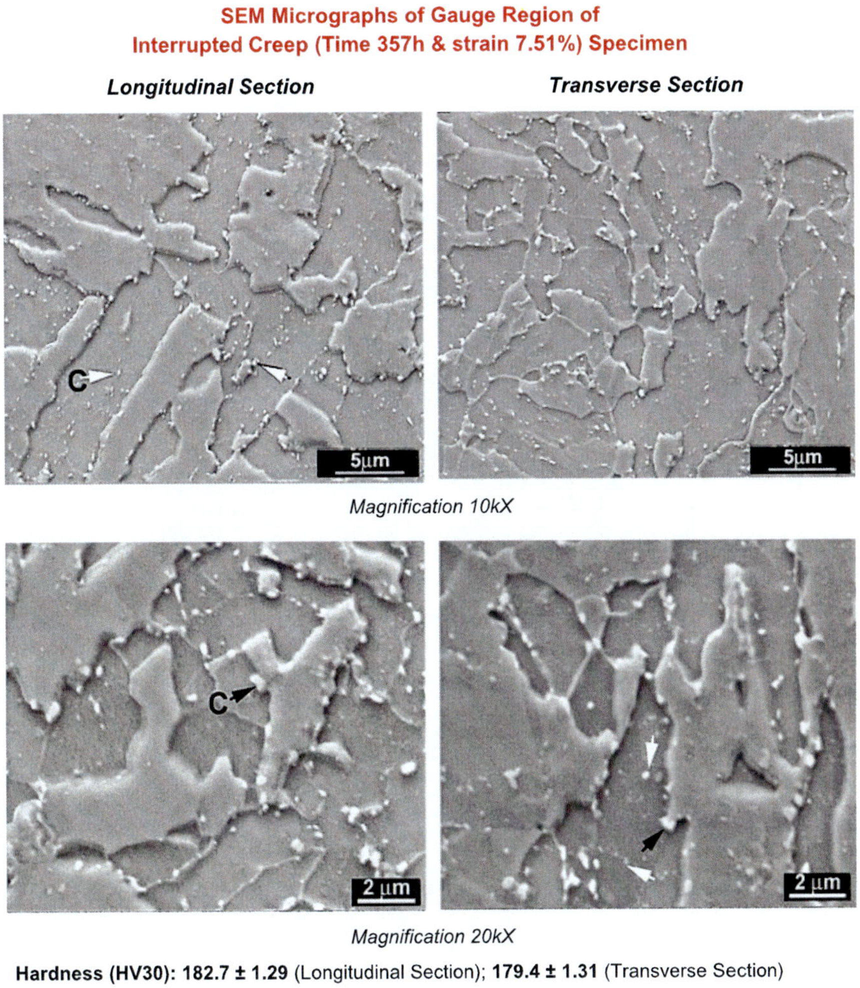

Fig. 4.70 SEM-SE micrographs of **longitudinal and transverse sections** of P91 (9Cr-1Mo) steel after creep exposed at 650 °C/100 MPa for 357 h, showing the distribution of fine carbides (white arrows) at grain interiors and coarse carbides (black arrows) at grain boundaries

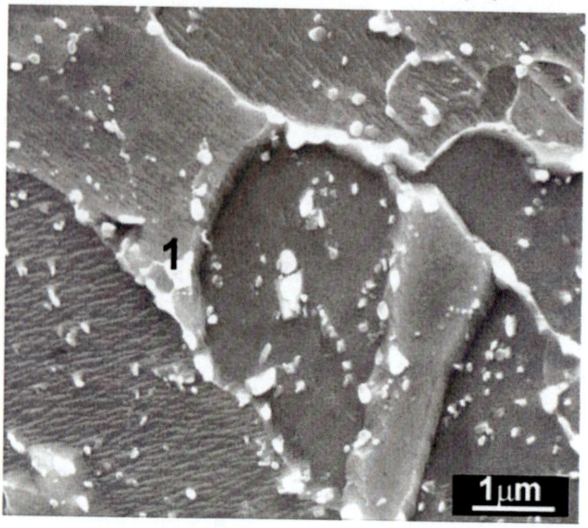

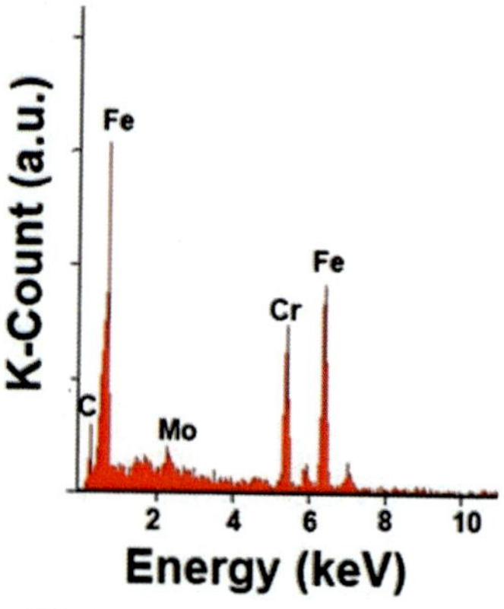

Fig. 4.71 SEM-SE micrograph (magnification 40kX) of ***longitudinal section*** of P91 (9Cr-1Mo) steel after creep exposed at 650 °C/100 MPa for 357 h, showing Cr-rich $M_{23}C_6$ [M = Fe, Cr, Mo] (point 1) coarse precipitates at grain interiors and grain boundaries of martensitic matrix

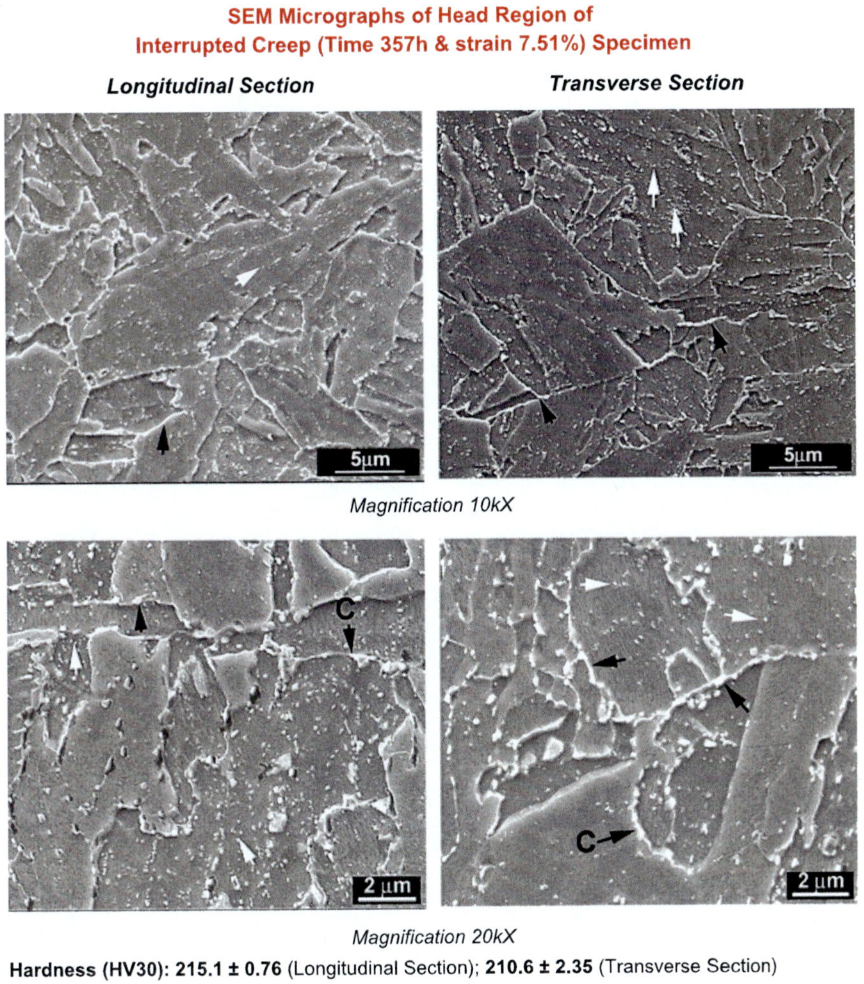

Fig. 4.72 SEM-SE micrographs of **longitudinal and transverse sections** of P91 (9Cr-1Mo) steel after creep exposed at 650 °C/100 MPa for 357 h, showing the distribution of coarse (black arrows) carbides (C) at grain boundaries and fine (white arrows) in grain interiors of martensitic matrix

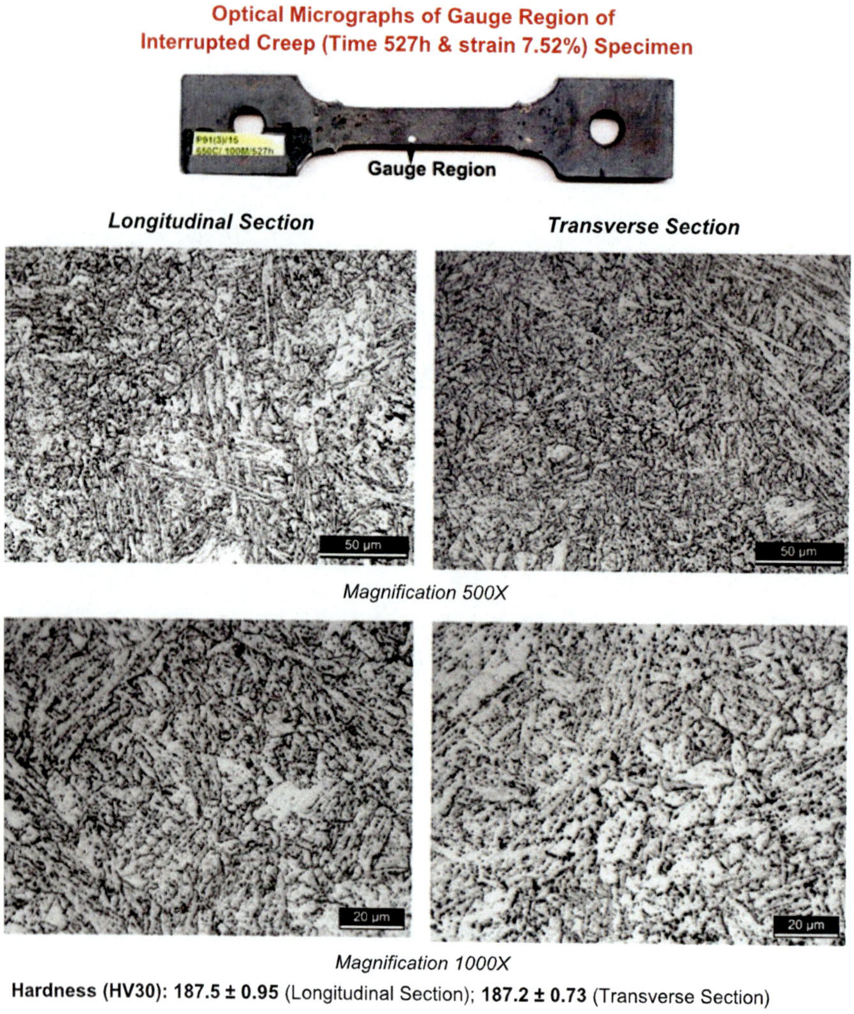

Fig. 4.73 Optical micrographs of **longitudinal and transverse sections** of P91 (9Cr-1Mo) steel after creep exposed at 650 °C/100 MPa for 527 h, explaining structure martensite

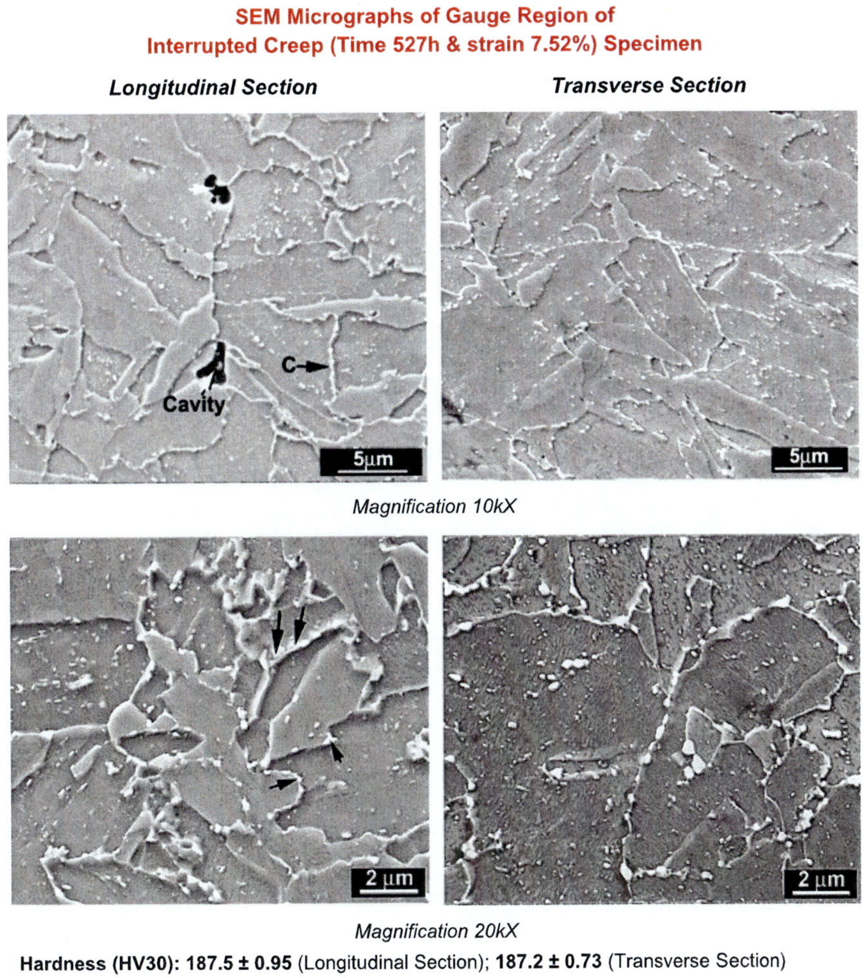

Fig. 4.74 SEM-SE micrographs of **longitudinal and transverse sections** of P91 (9Cr-1Mo) steel after creep exposed at 650 °C/100 MPa for 527 h, showing the distribution of (Fe, Cr, Mo)-rich carbides (C, black arrows) at grain boundaries and cavities (white arrows) in lath, and blocky martensitic structures

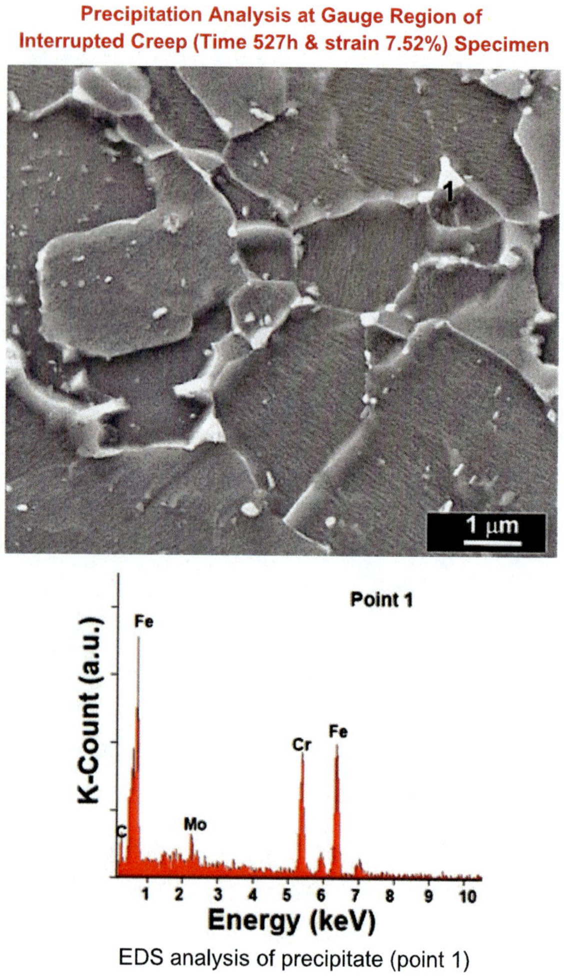

Fig. 4.75 SEM-SE micrograph (magnification 30kX) of ***longitudinal section*** of P91 (9Cr-1Mo) steel after creep exposed at 650 °C/100 MPa for 527 h, showing Cr-rich $M_{23}C_6$ [M = Fe, Cr, Mo] (point 1) precipitates at triple point of grain boundaries

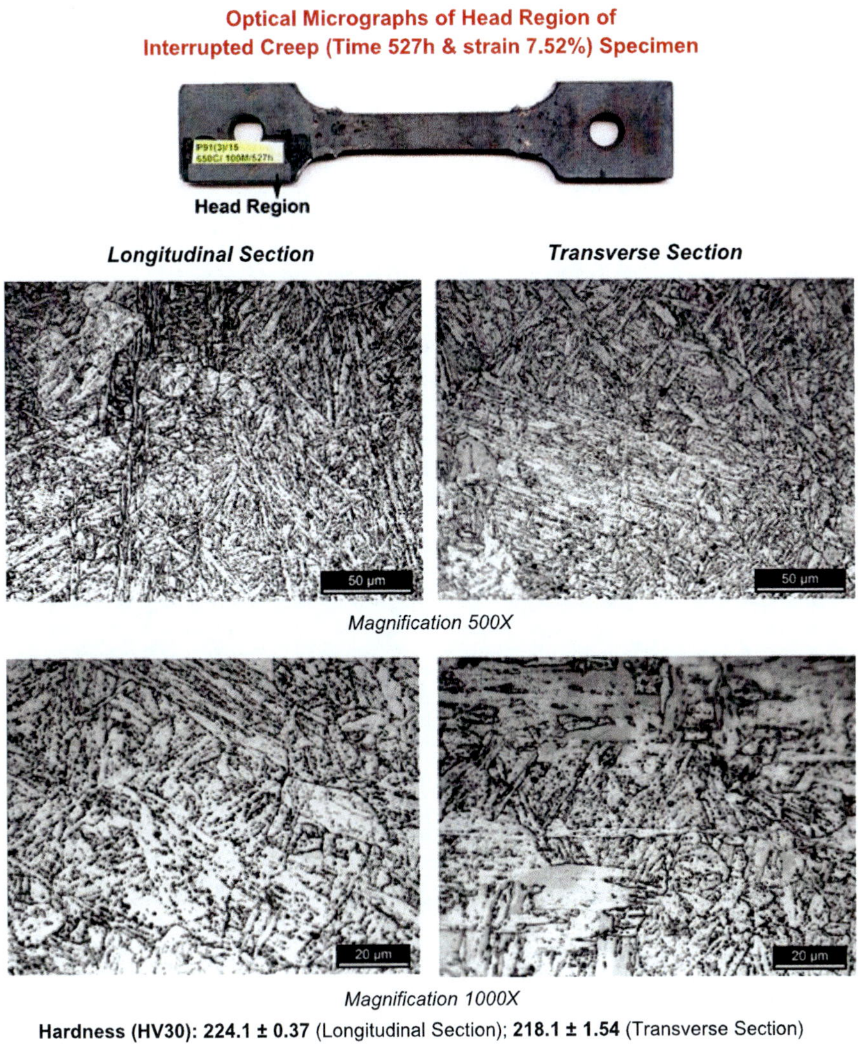

Fig. 4.76 Optical micrographs of **longitudinal and transverse sections** of P91 (9Cr-1Mo) steel after creep exposed at 650 °C/100 MPa for 527 h, explaining martensitic structures

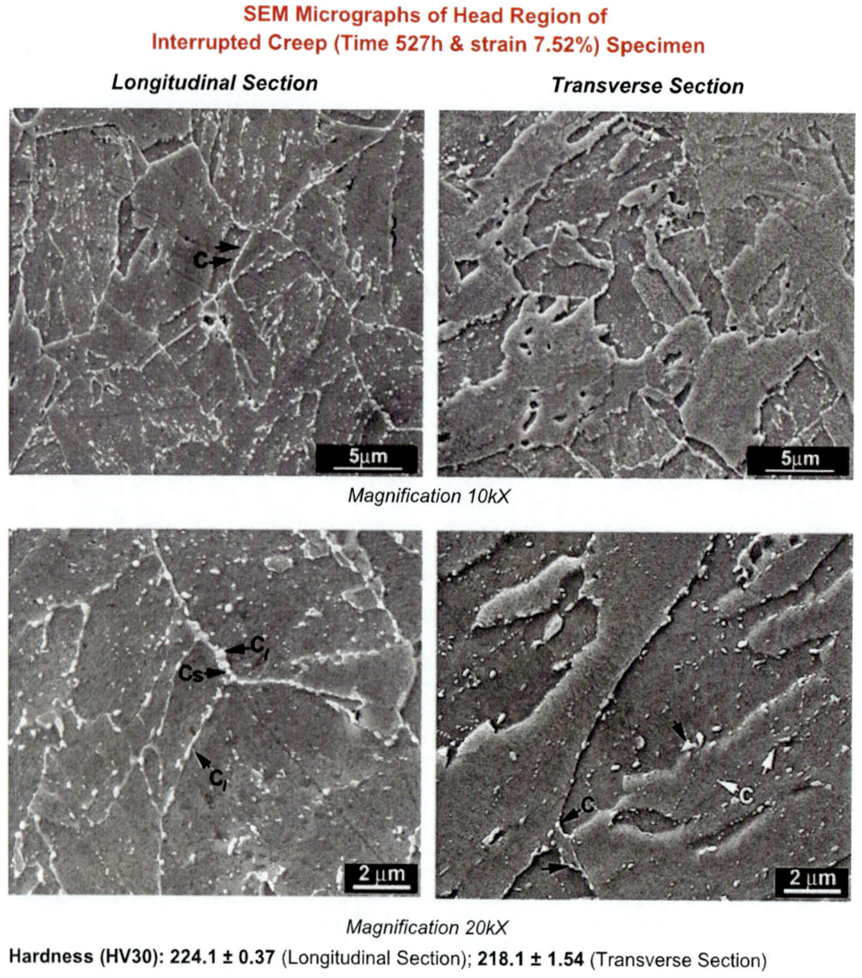

Fig. 4.77 SEM-SE micrographs of **longitudinal and transverse sections** of P91 (9Cr-1Mo) steel after creep exposed at 650 °C/100 MPa for 527 h, explaining the distribution of lenticular (Cl) and spherical (Cs) shaped fine carbides (C, white arrows) in grain interiors and coarse carbides (C, black arrows) at grain boundaries of martensitic matrix

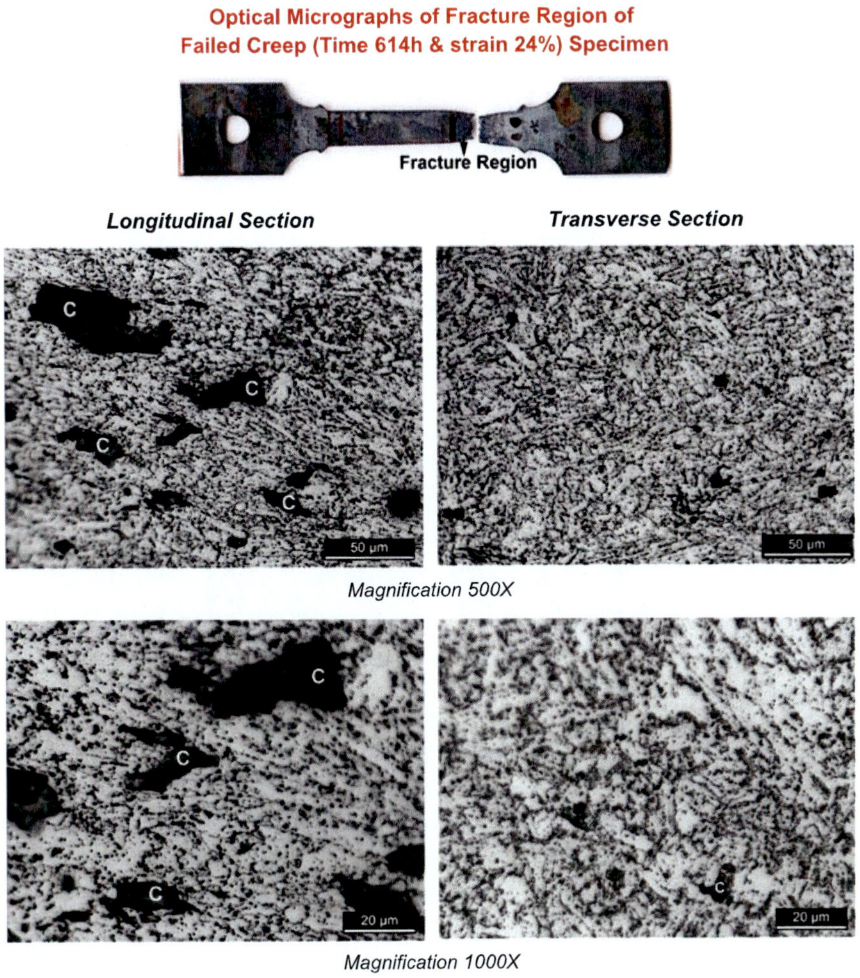

Fig. 4.78 Optical micrographs of **longitudinal and transverse sections** of P91 (9Cr-1Mo) steel after creep exposed at 650 °C/100 MPa for 614 h, showing the distribution of large size cavities (C) in the martensitic matrix

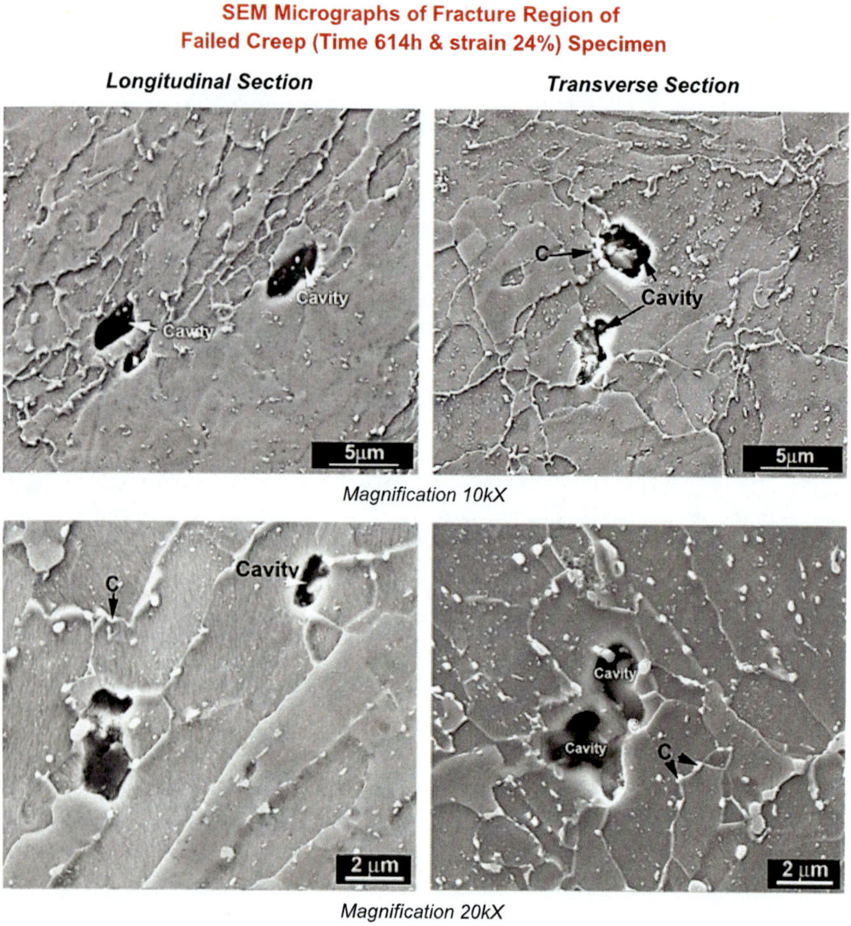

Fig. 4.79 SEM-SE micrographs of **longitudinal and transverse sections** of P91 (9Cr-1Mo) steel after creep exposed at 650 °C/100 MPa for 614 h, explaining the distribution of cavities and coarse carbides (C) along grain boundaries and fine carbides in grain interiors

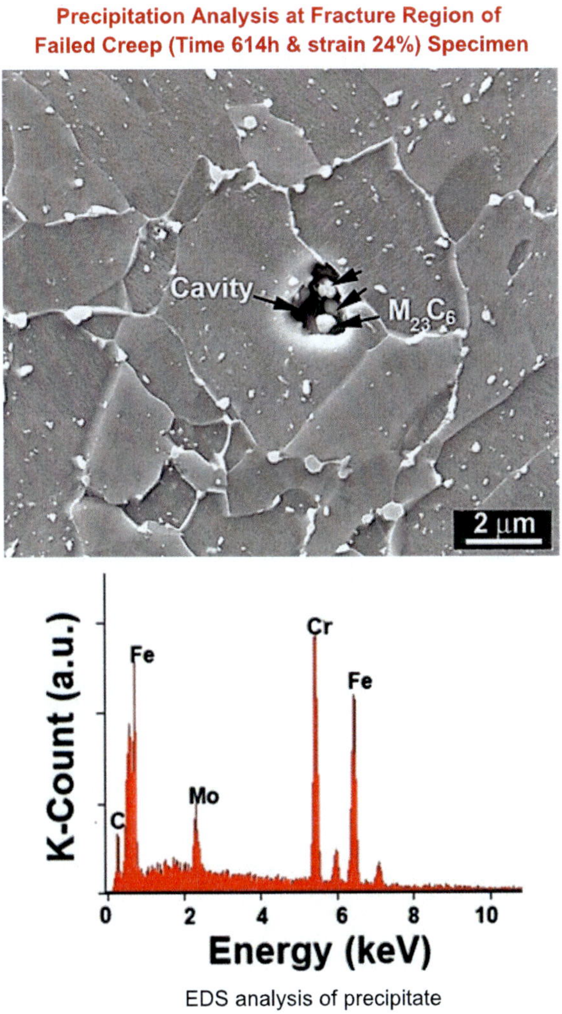

Fig. 4.80 SEM-SE micrograph (magnification 20kX) of ***transverse sections*** of P91 (9Cr-1Mo) steel after creep exposed at 650 °C/100 MPa for 614 h, showing the cavities along with Cr-rich $M_{23}C_6$ [M = Fe, Cr, Mo] precipitates at inside of cavities

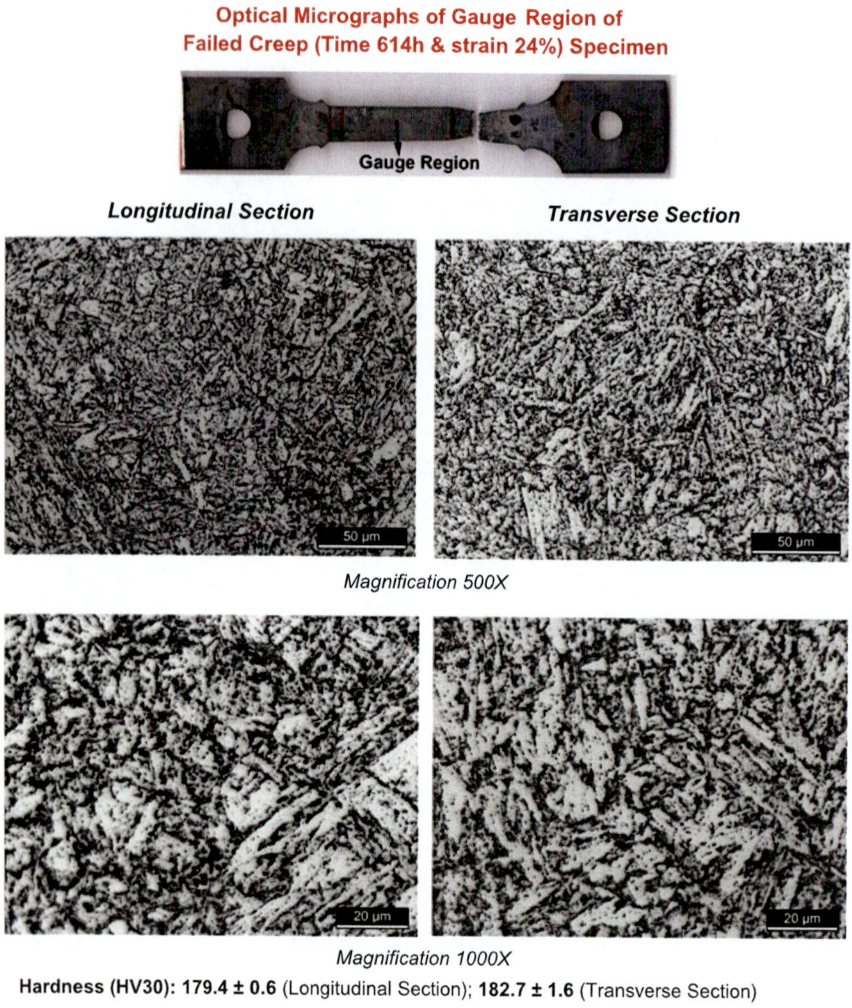

Fig. 4.81 Optical micrographs of **longitudinal and transverse sections** of P91 (9Cr-1Mo) steel after creep exposed at 650 °C/100 MPa for 614 h, showing martensitic structures

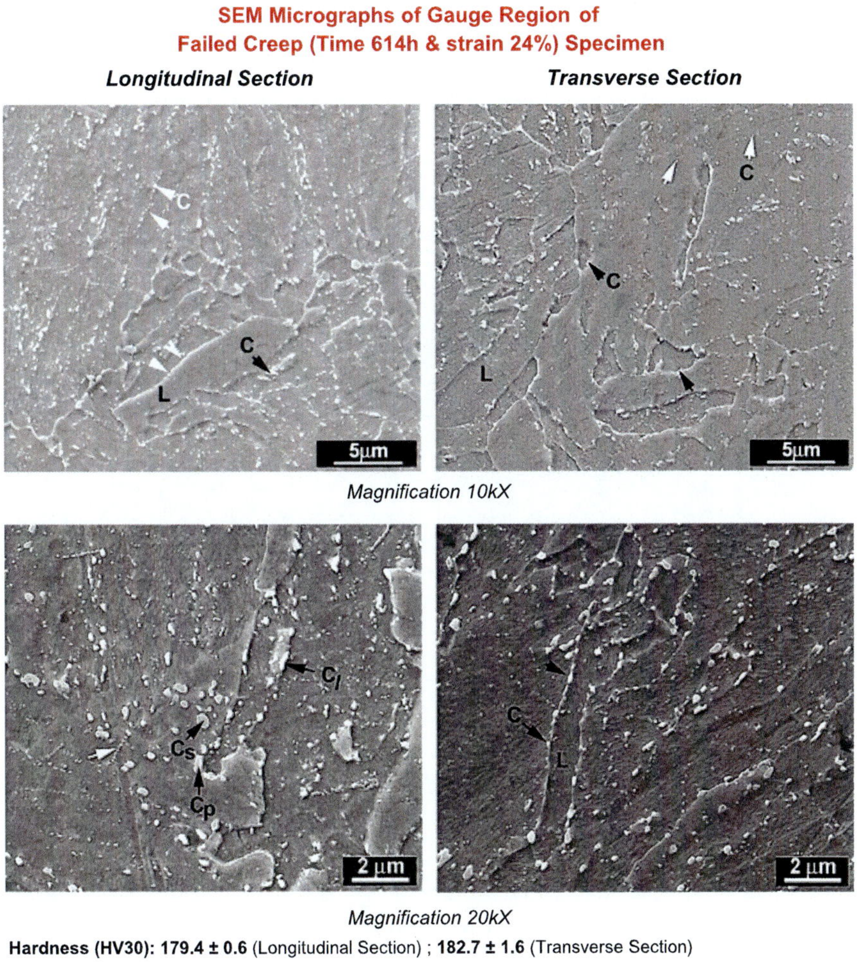

Fig. 4.82 SEM-SE micrographs of **longitudinal and transverse sections** of P91 (9Cr-1Mo) steel after creep exposed at 650 °C/100 MPa for 614 h, showing coarse carbides (black arrows) of different shapes, lenticular (C_l), spherical (C_s) and parallelepiped (C_p) and fine carbides (white arrows) along lath (L) boundaries and grain interiors

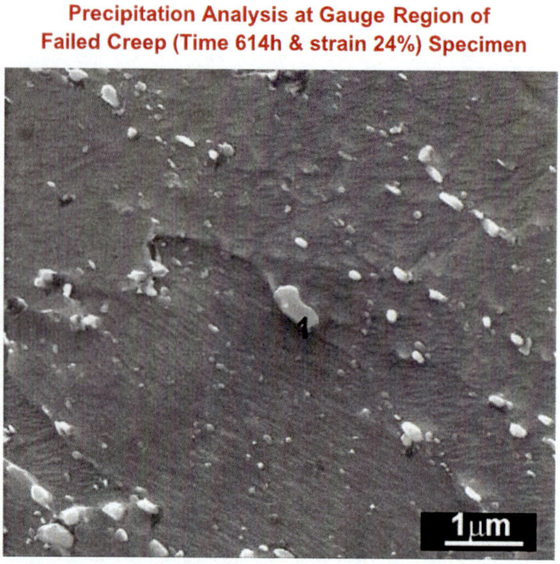

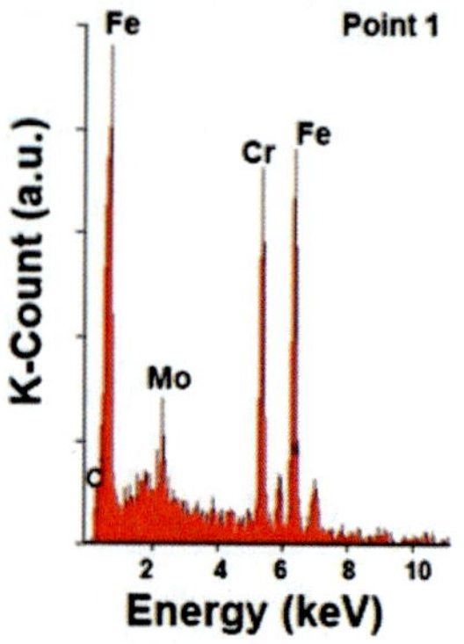

EDS analysis of precipitate

Fig. 4.83 SEM-SE micrograph (magnification 40kX) of ***transverse section*** of P91 (9Cr-1Mo) steel after creep exposed at 650 °C/100 MPa for 614 h, showing the distributions of Cr-rich $M_{23}C_6$ [M = Fe, Cr, Mo] precipitates of parallelepiped (point 1) and spherical shaped at martensite matrix

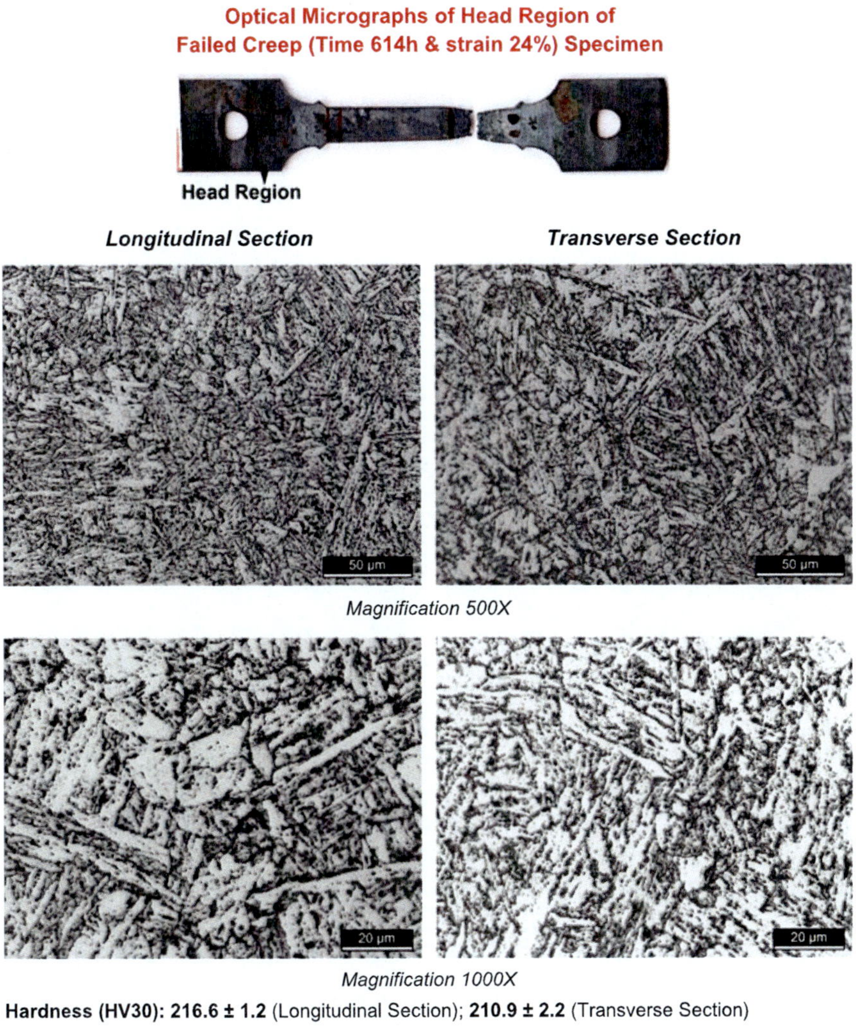

Fig. 4.84 Optical micrographs of **longitudinal and transverse sections** of P91 (9Cr-1Mo) steel after creep exposed at 650 °C/100 MPa for 614 h, showing martensitic structures

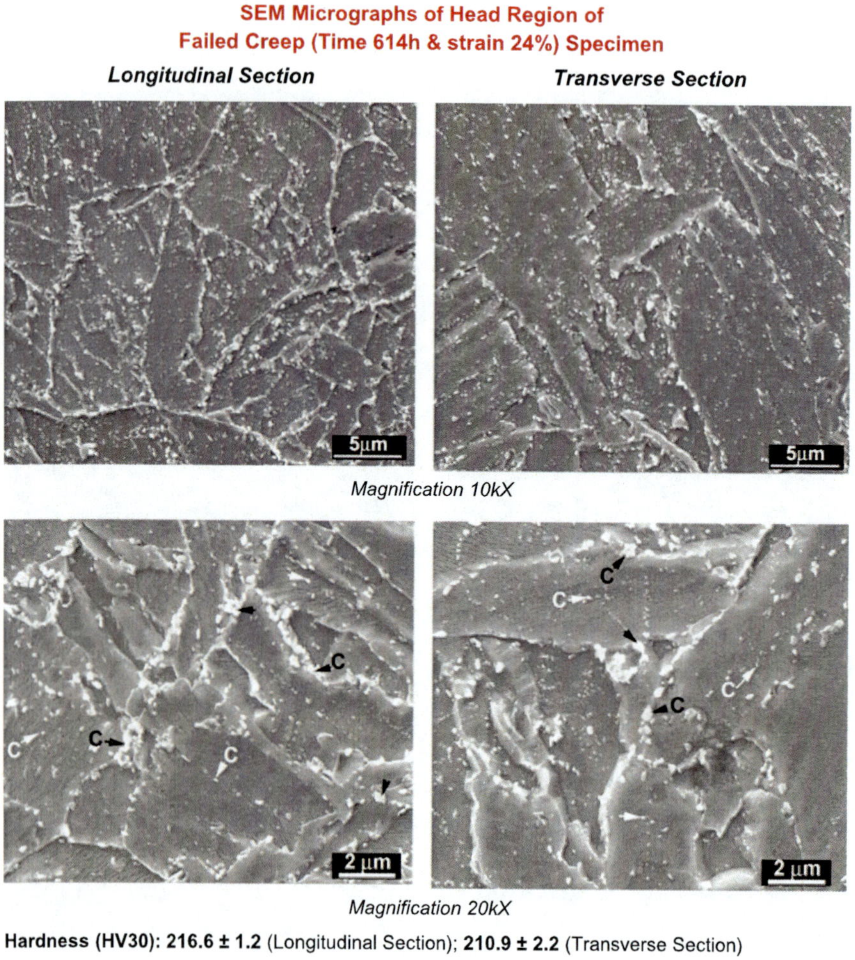

Fig. 4.85 SEM-SE micrographs of **longitudinal and transverse sections** of P91 (9Cr-1Mo) steel after creep exposed at 650 °C/100 MPa for 614 h, showing the distribution of coarse (black arrows) and fine (white arrows) carbides (C) at martensitic matrix

Chapter 5
Effect of Creep Exposure on Material Behaviors

5.1 Precipitate Size Variation

Upon creep progressing at primary stage, fine $M_{23}C_6$ carbide generation leads to increase of area fraction and decrease of mean diameter. Laves phase formation at higher creep exposure ($\varepsilon_c = 2.35\%$, 2010 h) increasing mean diameter, but not much enhance of area fraction. Higher creep strain ($\varepsilon_c = 3.28\%$, $t = 4746$ h) results in the enhancement of mean diameter and area fraction due to coalescence of precipitate (Fig. 4.19). At failure ($\varepsilon_c = 15\%$, $t = 6808$ h), dissolution or dragging out of particles from matrix at high strain results in decrease of size and area fraction of precipitates (Fig. 5.1).

5.2 Hardness Variation

Figures 5.2 and 5.3.

5.3 Magnetic Parameter Variation

With increasing creep strain up to 3.5%, the Multiple Samples and Single Interruption (MSSI) show samples show magnetic hardening with increase in coercivity and decrease in rms voltage. Due to formation of Cr-rich carbides and Laves phase at different creep stages, the magnetic domain movement is hindered and steel becomes magnetically harder with simultaneous coalescence and growth of precipitates (as explained in Fig. 5.1). Beyond 3.5% creep strain, the material shows the drop in coercivity and MBE rise, indicating magnetic softening due to dissolution or dragging out of particles from matrix at high strain as the material proceeded toward failure (Fig. 5.4).

R. K. Roy et al., *Microstructure Atlas of P91 Steel*,
https://doi.org/10.1007/978-981-19-9562-0_5

The Single Sample under Multiple Interruption (SSMI) sample shows continuous rise in coercivity and drop in MBE rms voltage upto 4000 h creep time due to magnetic hardening with simultaneous effect of precipitation coarsening and coalescence. Beyond 4000 h, the material showed drop in coercivity (MBE rise), indicating magnetic softening due to dissolution or dragging out of particles from matrix at high strain as the material proceeded toward failure (Fig. 5.5).

With increasing creep strain up to 3%, the material P91 shows magnetic hardening with increase in coercivity and decrease in rms voltage due to precipitation of finer Cr-rich carbides. Beyond 3% creep strain, the simultaneous formation of Laves phase and finer Cr-rich carbide do not have much effect on coercivity and MBE voltage. Beyond 6% creep strain, the material again showed magnetic hardening (coercivity rise, MBE drop) due to carbide coarsening. The large-sized carbides' restrict magnetic domain wall movement, leading to such magnetic hardening (Fig. 5.6).

With increasing creep strain up to 4–6%, the material shows magnetic softening with decrease in coercivity. This rapid magnetic softening may be attributed to fine carbide precipitation. However, the MBE value also shows a decrease in view of restricted response toward such morphological changes in the bulk owing to low penetration depth of MBE. Beyond 4% creep strain, there is a rapid magnetic hardening (increased coercivity) due to growth and simultaneous coalescence of carbides occurring at increased creep strain. The dissolution or dragging out of particles from matrix at high strain (>12%) results in magnetic softening, decreasing coercivity, and increasing MBE voltage (Fig. 5.7).

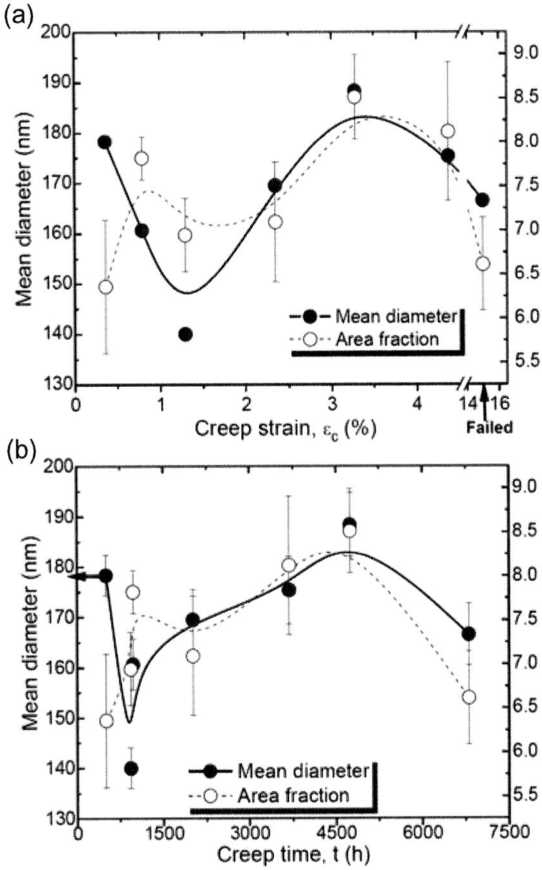

Fig. 5.1 Variation of mean diameter and area fraction of carbide precipitates in P91 steel under creep exposed at 550 °C/200 MPa as a function of **a** creep strain and **b** creep time

Fig. 5.2 Hardness variation in gauge regions of P91 steels under creep exposed of 550 °C/200 MPa as a function of **a** creep time and **b** strain

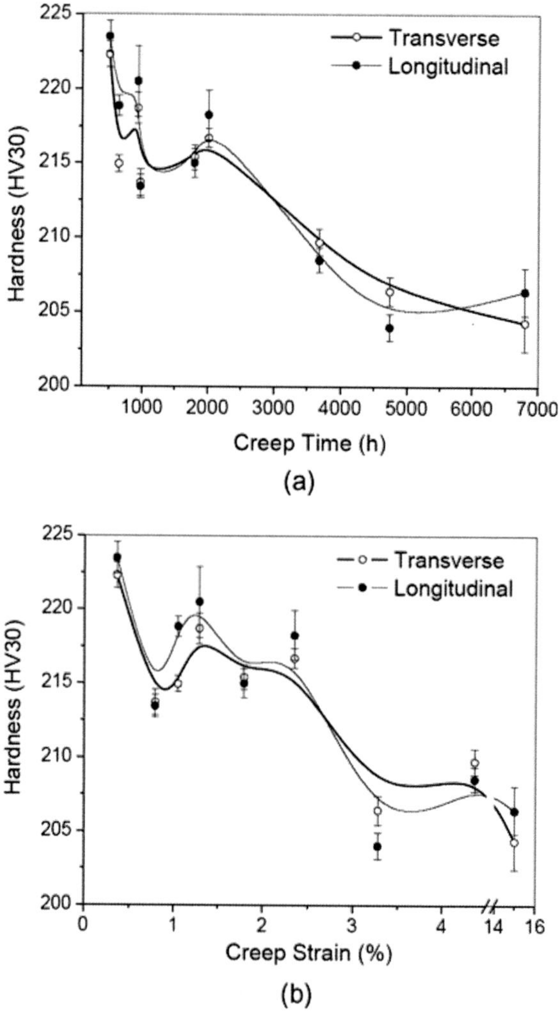

Fig. 5.3 Hardness variation in head regions of P91 steel under creep exposed of 550 °C/200 MPa as a function of **a** creep time and **b** strain

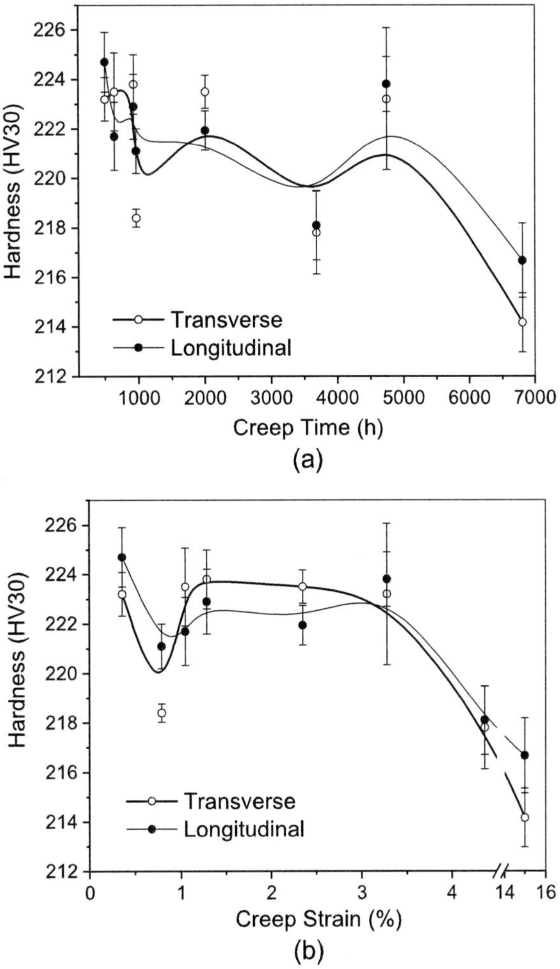

Fig. 5.4 Creep strain effect on coercivity and MBE voltage of P91 steel for ***M***ultiple ***S***amples and ***S***ingle ***I***nterruption (***MSSI***) under creep condition of 550 °C/200 MPa

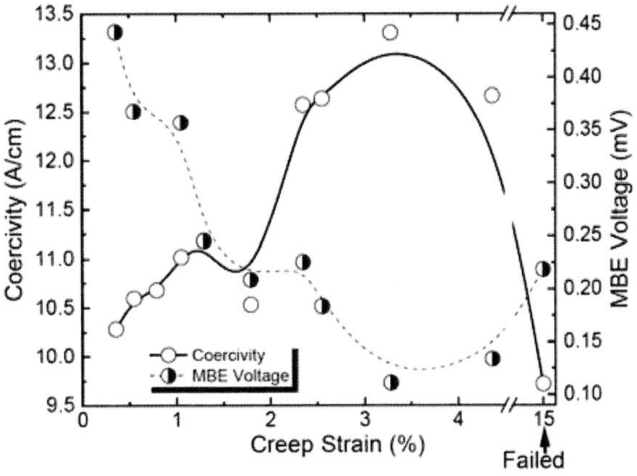

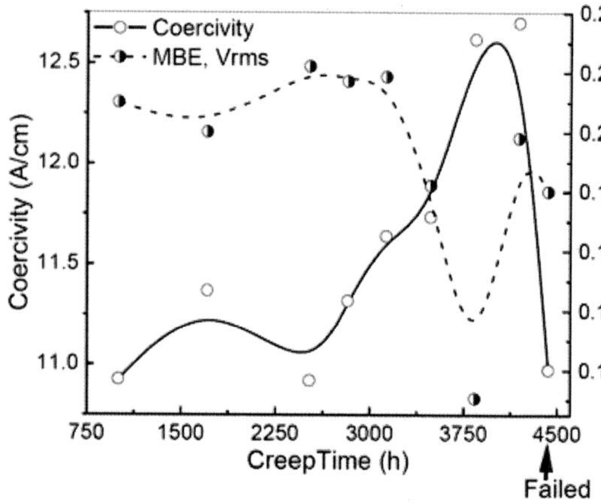

Fig. 5.5 Variation of coercivity and MBE rms voltage of P91 steel for ***S***ingle ***S***ample under ***M***ultiple ***I***nterruption (***SSMI***) under creep condition of 550 °C/200 MPa as a function of creep time

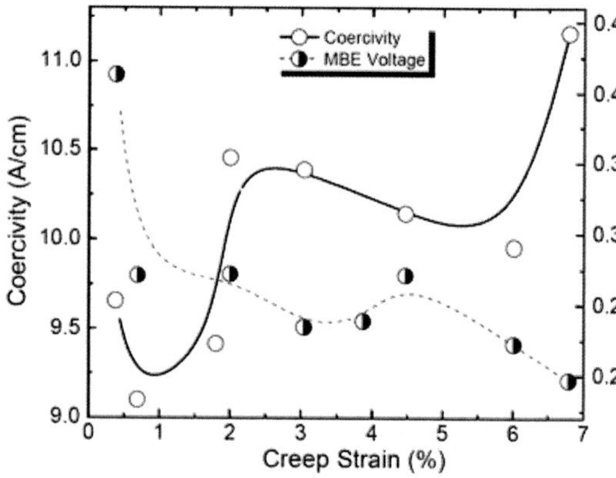

Fig. 5.6 Creep strain effect on coercivity and MBE voltage of P91 steel for ***M***ultiple ***S***amples and ***S***ingle ***I***nterruption (***MSSI***) under creep condition of 600 °C/140 MPa

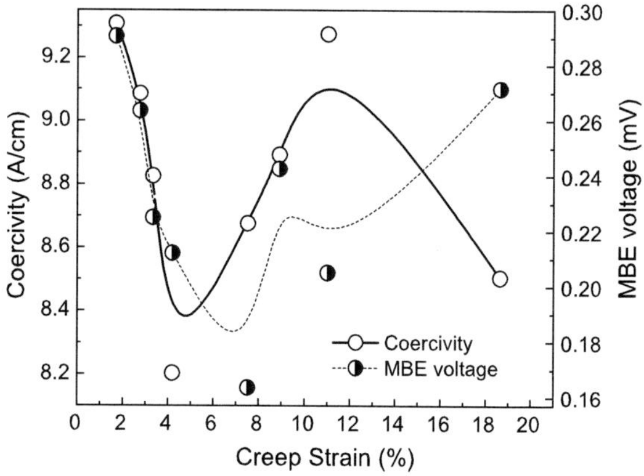

Fig. 5.7 Creep strain effect on coercivity and MBE voltage of P91 steel for ***M***ultiple ***S***amples and ***S***ingle ***I***nterruption (***MSSI***) under creep condition of 650 °C/100 MPa

5.4 Nonlinear Ultrasonic (NLU) Parameter Variation

The variation of NLU parameter (β) with creep strain follows the trend of change in size and volume fraction of precipitates as a function of creep (as explained in Fig. 5.1). Initial increase is due to the dislocations and increase in number of precipitates. No change of NLU parameter (β) in the range of 2–4% strain is because of no such change in volume fraction of precipitates. A sharp increase of β above 4% creep strain is observed as the precipitates grow and diameter of precipitates is beyond the coherency limit (Fig. 5.8).

Initial increase of NLU parameter (β) is due to the dislocations and increase in number of precipitates. No change of β in creep time of 1000–3500 h is due to no such change in volume fraction of precipitates. A sharp increase of β above 3500 h is observed as the precipitate grows and diameter of precipitates is beyond the coherency limit (Fig. 5.9).

Initial increase of NLU parameter (β) is due to the dislocations and increase in number of precipitates. It is followed by no change of NLU parameter (β) due to no such change in volume fraction of precipitates. A sharp increase of β above 4% creep strain is observed as the precipitates grow and diameter of precipitates is beyond the coherency limit (Fig. 5.10).

At initial creep strain, a little change of NLU parameter (β) is due to small change in volume fraction of precipitates. A sharp increase of β at higher creep strain is observed as the precipitates grow and diameter of precipitates is beyond the coherency limit (Fig. 5.11).

Fig. 5.8 Creep strain effect on NLU parameter of P91 steel for ***M***ultiple ***S***amples and ***S***ingle ***I***nterruption (***MSSI***) under creep condition of 550 °C/200 MPa

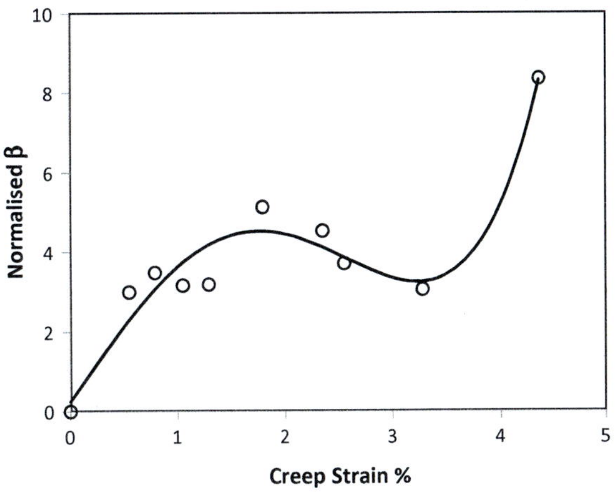

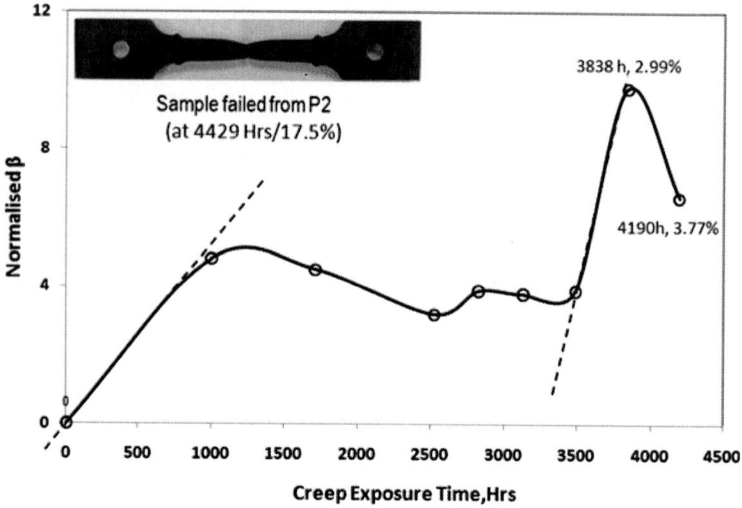

Fig. 5.9 Variation of NLU parameter of P91 for ***S***ingle ***S***ample under ***M***ultiple ***I***nterruption (***SSMI***) under creep condition of 550 °C/200 MPa as a function of creep time

Fig. 5.10 Creep strain effect on NLU parameter of P91 steel for ***M***ultiple ***S***amples and ***S***ingle ***I***nterruption (***MSSI***) under creep condition of 600 °C/140 MPa

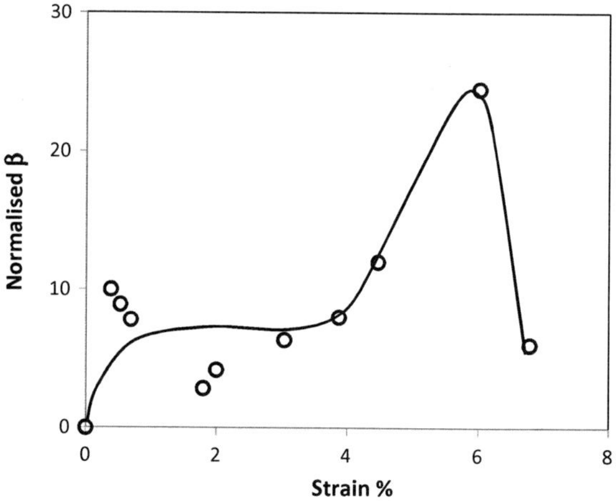

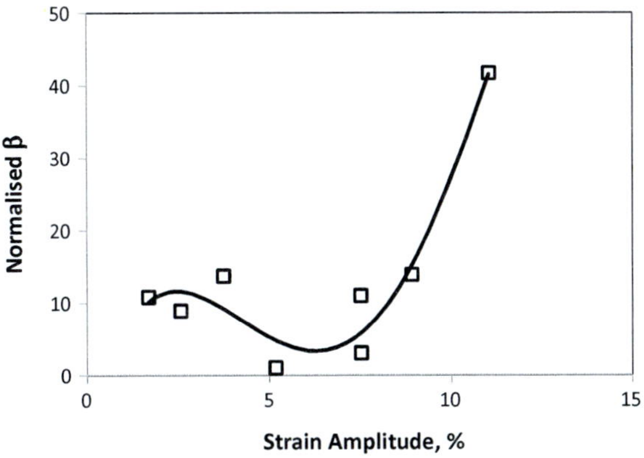

Fig. 5.11 Creep strain effect on NLU parameter of P91 steel for ***M***ultiple ***S***amples and ***S***ingle ***I***nterruption (***MSSI***) under creep condition of 650 °C/100 MPa

5.5 Conventional Ultrasonic Parameter Variation

The ultrasonic attenuation, which is the sum of absorption and scattering, is mainly dependent upon the damping capacity and the scattering from the grain boundary, wedge cracks (if present), etc. in the material. The variation of attenuation with creep strain follows the trend of change in size and volume fraction of precipitates as a function of creep (as explained in Fig. 5.1). Initial increase of attenuation is due to the dislocations and increase in number of precipitates. A little change of attenuation in the range of 1–3% strain is due to no such change in volume fraction of precipitates. A sharp increase of attenuation above 3% creep strain is observed due to growth and simultaneous coalescence of carbides occurring at increased creep strain (Fig. 5.12).

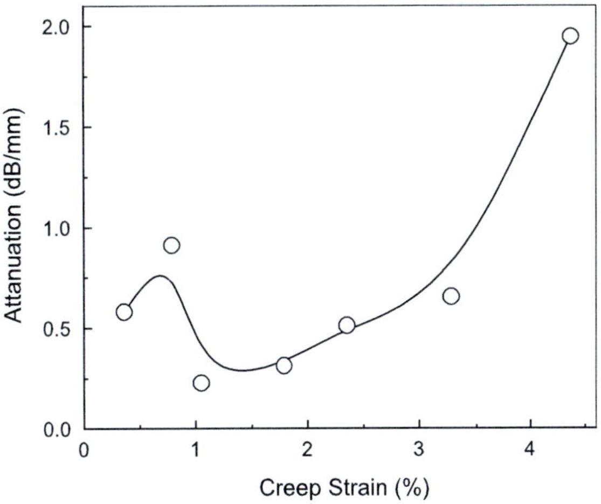

Fig. 5.12 Creep strain effect on ultrasonic parameter of P91 steel for ***M***ultiple ***S***amples and ***S***ingle ***I***nterruption (***MSSI***) under creep condition of 550 °C/200 MPa

Initial increase of ultrasonic attenuation is due to the dislocations and increase in number of precipitates. A little change of attenuation above 2% strain is due to no such change in volume fraction of precipitates (Fig. 5.13).

Initial increase of ultrasonic attenuation is due to the dislocations and increase in number of precipitates. A small change of attenuation in the range of 4–12% creep strain is due to no such change in volume fraction of precipitates. An increase of attenuation beyond 12% creep strain is observed due to growth and simultaneous coalescence of carbides occurring at increased creep strain (Fig. 5.14).

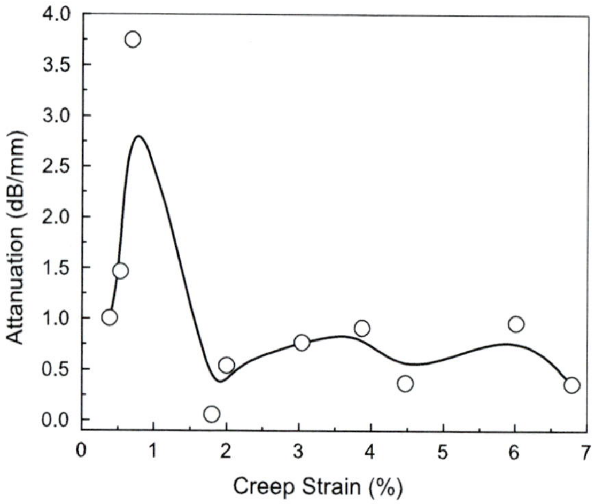

Fig. 5.13 Creep strain effect on ultrasonic parameter of P91 steel for ***M***ultiple ***S***amples and ***S***ingle ***I***nterruption (***MSSI***) under creep condition of 600 °C/140 MPa

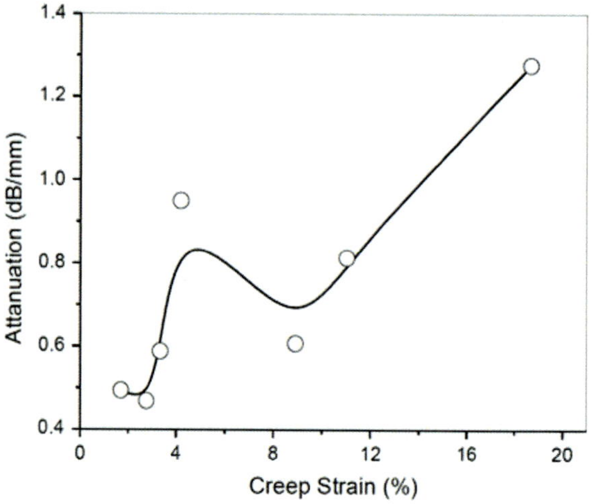

Fig. 5.14 Creep strain effect on ultrasonic parameter of P91 steel for ***M***ultiple ***S***amples and ***S***ingle ***I***nterruption (***MSSI***) under creep condition of 650 °C/100 MPa

Supplements

Brief Description of Experimental Methodologies

Sample Preparation

In this investigation, the creep specimen prepared along the longitudinal of pipe and its dimensions are as per ASTM standard (Fig. A.1).

Creep Testing

The creep test was carried out with standard sample (Fig. A.1) in the constant load stress rupture test machine (Mayes make, model-MT3, 30KN (max), 950 °C (max.)), as shown in Fig. A.2. Two R-type thermocouples were tied on the specimen at the end of gauge length section. A preload of 5–10% of the target load was used and the samples were heated to the test set temperature within 1-h duration and soaked for 30 min for temperature equalization along the gauge length. The test temperature was controlled within ±2 °C of the set temperature. Rupture time was noted from

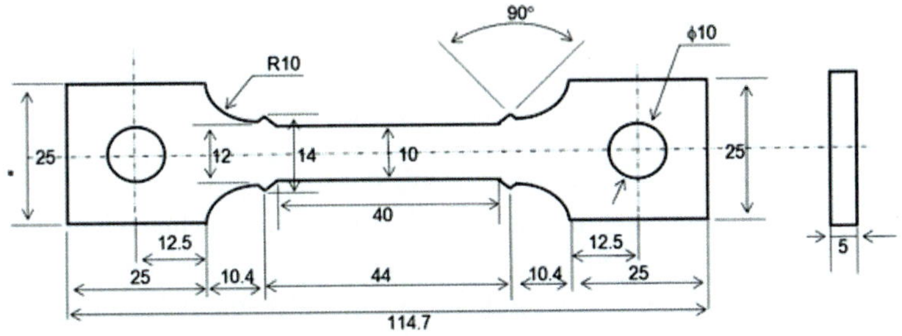

Fig. A.1 Creep testing specimen details. Dimensions (in mm)

R. K. Roy et al., *Microstructure Atlas of P91 Steel*,
https://doi.org/10.1007/978-981-19-9562-0

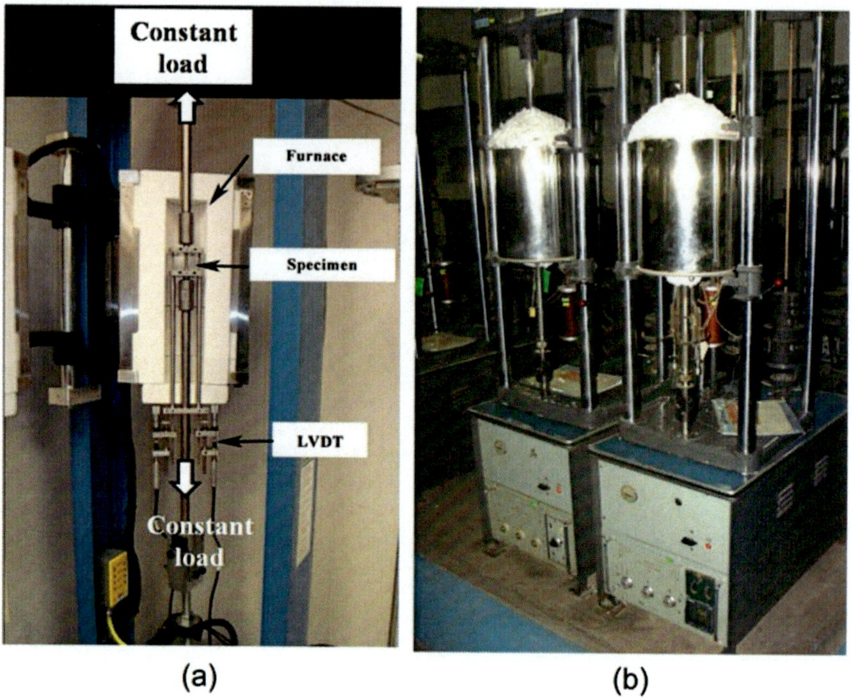

Fig. A.2 Creep testing machine. **a** Sample before put into furnace, **b** sample packed in the furnace

the elapsed hour meter attached to the test machine. The displacement was measured using a pair of LVDT attached with the extensometer. Initial displacement readings were manually recorded during hot-loading and thereafter continuous recordings were done at every 15-min interval using a data-logger.

Microscopy of Samples

Optical and Scanning Electron Microscopy

The samples of size $10 \times 5 \times 5$ mm^3 were cut along longitudinal and transverse (Fig. 2.1b) of pipe for as-received specimens and that of creep-exposed specimens at gauge and failed regions, as per requirement (Fig. A.3).

These specimens were suitably mounted by conductive mounting powders and were ground using 200–1200 grit silicon carbide abrasive papers. It was followed by coarse polishing on 10 μm cloths using 1 μm alumina solution and final polishing on 1 μm cloths using 0.05 μm alumina solution. The chemical etching was done by Vilella's reagent (1 g picric acid, 5 ml HCl, and 100 ml ethanol). Chemically

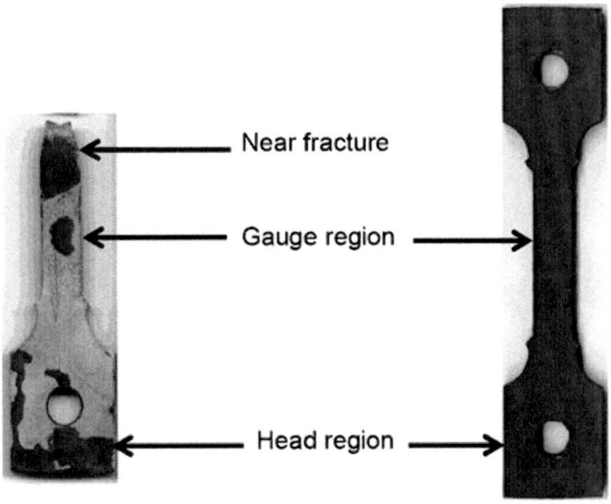

Fig. A.3 Microscopy of regions of creep-exposed specimens. **a** Creep failed and **b** creep interrupted Multiple Samples and Single Interruption (MSSI)

etched and dried specimens were examined under an optical microscope of Leica (model: DM2500M) and scanning electron microscope of FEI FEG-SEM (model: Nova NanoSEM430).

Optical microscopy (OM) was done under 200X, 500X, and 1000X magnifications, and the images were captured by a digital camera. The selective images with high resolution are presented in this microstructure Atlas.

Scanning electron microscopy (SEM) was operated at 15 kV and scanned at the lower magnification in order to extract the most representative microstructural features. The representative images are digitally captured at the magnifications of 5kX, 10kX, and 20kX. The compositional elemental analysis of precipitates was done under 15 kV voltage through EDS analyzer attached with SEM.

Transmission Electron Microscopy

For transmission electron microscopy (TEM), the thin specimen of 0.3 mm thickness was cut from as-received and creep-exposed samples by electric discharge machining (EDM) with a proper cooling arrangement. These specimens were further thinned down by 1200–2500 grit silicon carbide abrasive papers till 100 μm fine foil which was finally punched to 3 mm diameter disk. The disk sample is again thinned down by twin jet polishing using solution of 4% perchloric acid in methanol at −40 °C and making the tiny holes, enabling the hole adjacent area to be electron transparent.

A Philips (model: CM200) analytical transmission electron microscope operating at an accelerated voltage of 200 kV was used to identify carbide precipitates of as-received and creep-exposed samples. The precipitates were identified through selected area electron diffraction pattern (SAED) and elemental analyses by EDAX genesis attached with TEM.

Image Analysis

The microstructural images acquired by SEM and TEM were analyzed for the determination of size and area fraction of carbide precipitates by an image analysis software of Leica QWIN. Each data point of precipitate size and volume fraction was about the average of 1000 numbers of particles from 5 numbers of SEM images. The precipitate size is also confirmed by TEM.

Mechanical Property Determination

The tensile properties of as-received steel are determined by universal tensile testing machine under the crosshead speed of 5 mm/min at room and elevated temperatures of 550, 600, and 650 °C.

The hardness measurement of as-received and creep-exposed specimens was carried out by a diamond pyramid Vickers hardness tester of Chennai Metco (Economet VH50MD) at 30kgf load with a dwell time of 10 s for indenting. The sample surface was ground up to 1200 grit SiC paper before hardness measurement. Each hardness value was the average of 5–10 numbers of data points.

Brief Description of NDE Techniques

Electromagnetic Method

Among the electromagnetic techniques, Magnetic Barkhausen Emissions (MBE) and Magnetic Hysteresis Loop (MHL) have been used under this project. The magnetic hysteresis loop and the magnetic Barkhausen emission signal are shown schematically in Fig. A.4 along with the measuring parameters. However, the coercivity (Hc) and the rms voltage (Vrms) are the most important parameters as regards with the microstructural changes are concerned. Hence, the emphasis has been given to measure those parameters. Both MHL and MBE parameters are to be measured by an electromagnetic sensing device, named *MagStar* (Fig. A.5), and developed jointly by CSIR-NML and M/s, Technofour, Pune.

For MBE measurement, a sinusoidal cyclic magnetic field of 300 Oe with an excitation frequency of 40 Hz was induced, and the Barkhausen signals were filtered with a wide band-pass filter (30–300 kHz). On the other hand, a sinusoidal cyclic magnetic field of 1500 Oe with an excitation frequency of 50 MHz was induced for MHL measurement.

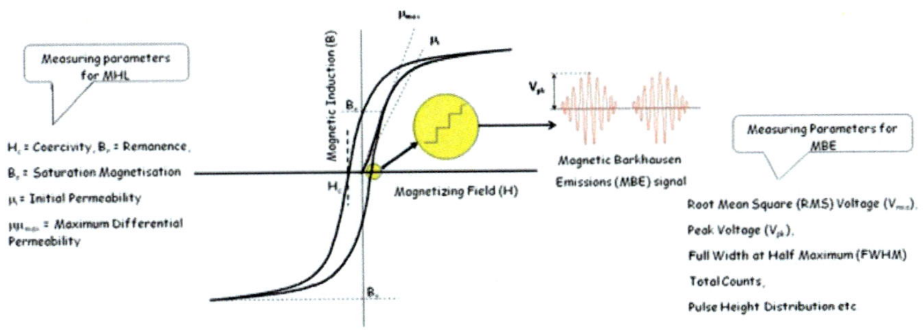

Fig. A.4 Schematic representation of Magnetic Hysteresis Loop (MHL) and Magnetic Barkhausen Emissions (MBE)

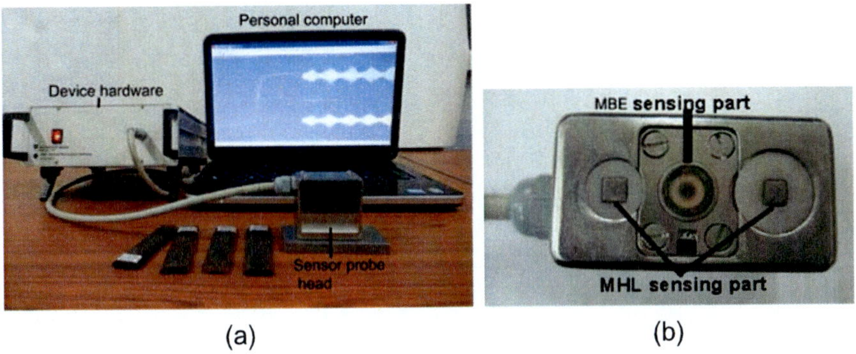

Fig. A.5 **a** *MagStar* device, **b** combo sensor probe head for MHL and MBE measurement

Nonlinear Ultrasonic (NLU) Technique

The variation in the distribution of discontinuities due to material degradation results in a variation of nonlinear elasticity, and material degradation can be monitored by nonlinear ultrasonic (NLU) measurement. The most well-known NLU response is second harmonic generation caused by higher order nonlinear elasticity of the materials in question. The relation between arbitrary constants C_1, C_2 of nonlinear elasticity and two amplitudes of distorted ultrasound, fundamental ($A1$) and second harmonic (A_2), was derived by Breazeal and Ford et al. using fluid analogy and Lagrange's equation of motion:

$$\beta = \frac{C_1}{C_2} = \frac{8}{k^2 x}\frac{A_2}{A_1^2} \tag{A.1}$$

where K and x are wave number and propagation distance, respectively. The right-hand side of Eq. (A.1), which represents the ratio of primary and second-order terms of nonlinear stress–strain relation, is referred to as ultrasonic nonlinear parameter β.

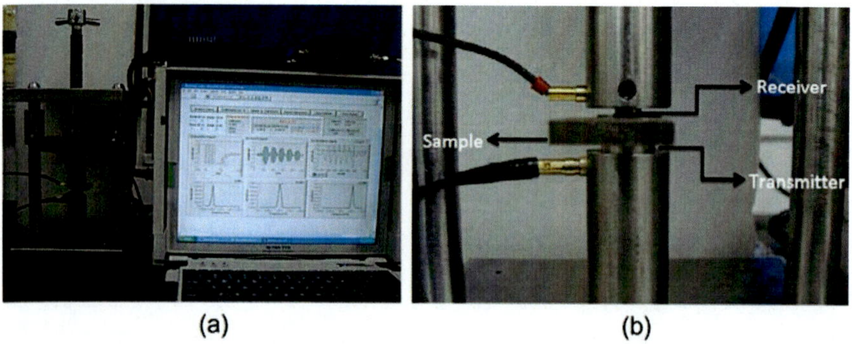

Fig. A.6 **a** "Ultraβ": Portable system for NLU parameter measurement, **b** sample holder for applying constant load to the sample

The effect of microstructure transition by degradation or material processing to β variation has been widely studied, i.e., precipitation hardening heat treatment, cold work, isothermal aging, creep, and fatigue, and has been reported that dislocation density, precipitation distribution, microvoid, and externally applied or residual stress are primary factors to ultrasonic nonlinear parameter β.

A portable instrument "Ultraβ" designed and developed at CSIR-NML was used to measure β (Fig. A.6a). A sample holder is also designed to place the sample in between the transmitter and receiver probes under constant load (Fig. A.6b).

For the measurement of NLU parameter, a 5 MHz narrowband transmitter and a 10 MHz broadband receiver were used. NLU measurements were carried out in each sample in the as-received condition prior to creep testing (β_{zero}) and after interruption (β_{damaged}). The normalized β was defined as mentioned in Eq. (A.2):

$$\beta = \frac{\beta_{\text{damaged}} - \beta_{\text{zero}}}{\beta_{\text{zero}}} \tag{A.2}$$

A 5-cycle tone burst signal was used to excite the transmitter at two different voltages, 1050 and 1200 V. The received signal was analyzed using *Calβ* © software for online NLU parameter determination after calibration.